Global Bioethanol

Global Bioethanol

Evolution, Risks, and Uncertainties

Edited By

Sergio Luiz Monteiro Salles-Filho
Luís Augusto Barbosa Cortez
José Maria Ferreira Jardim da Silveira
Sergio C. Trindade
Maria da Graça Derengowski Fonseca

AMSTERDAM • BOSTON • HEIDELBERG • LONDON
NEW YORK • OXFORD • PARIS • SAN DIEGO
SAN FRANCISCO • SINGAPORE • SYDNEY • TOKYO

Academic Press is an imprint of Elsevier

Academic Press is an imprint of Elsevier
125 London Wall, London EC2Y 5AS, UK
525 B Street, Suite 1800, San Diego, CA 92101-4495, USA
50 Hampshire Street, 5th Floor, Cambridge, MA 02139, USA
The Boulevard, Langford Lane, Kidlington, Oxford OX5 1GB, UK

Notices
Knowledge and best practice in this field are constantly changing. As new research and experience broaden our understanding, changes in research methods, professional practices, or medical treatment may become necessary.

Practitioners and researchers must always rely on their own experience and knowledge in evaluating and using any information, methods, compounds, or experiments described herein. In using such information or methods they should be mindful of their own safety and the safety of others, including parties for whom they have a professional responsibility.

To the fullest extent of the law, neither the Publisher nor the authors, contributors, or editors, assume any liability for any injury and/or damage to persons or property as a matter of products liability, negligence or otherwise, or from any use or operation of any methods, products, instructions, or ideas contained in the material herein.

British Library Cataloguing-in-Publication Data
A catalogue record for this book is available from the British Library

Library of Congress Cataloging-in-Publication Data
A catalog record for this book is available from the Library of Congress

ISBN: 978-0-12-803141-4

For Information on all Academic Press publications
visit our website at http://www.elsevier.com/

  Working together
to grow libraries in
developing countries

www.elsevier.com • www.bookaid.org

Publisher: Joe Hayton
Acquisition Editor: Raquel Zanol
Editorial Project Manager: Mariana Kühl Leme
Editorial Project Manager Intern: Ana Claudia Abad Garcia
Production Project Manager: Kiruthika Govindaraju
Designer: Maria Inês Cruz

Typeset by MPS Limited, Chennai, India

Contents

3. Political Orientations, State Regulation and Biofuels in the Context of the Food–Energy–Climate Change Trilemma

M. Harvey and Z.P. Bharucha

4. Innovation Systems of Ethanol in Brazil and the United States: Making a New Fuel Competitive

L.C. de Sousa, N.S. Vonortas, I.T. Santos and D.F. de Toledo Filho

8. Technological Foresight of the Bioethanol Case

José Maria F. J. da Silveira, M.E.S. Dal Poz, L.G. Antonio de Souza and I.R.L. Huamani

9. China's Fuel Ethanol Market

H. Lu

List of Contributors

L.G. Antonio de Souza Interdisciplinary Center of Energy Planning, State University of Campinas, Campinas, São Paulo, Brazil

Wilson A. Araújo DuPont Industrial Biosciences, Paulínia, Brazil

R. Baldassin, Jr. Interdisciplinary Center for Energy Planning – NIPE, UNICAMP, Campinas, São Paulo, Brazil

Z.P. Bharucha Department of Sociology, University of Essex, Colchester, United Kingdom

A. Bin School of Applied Sciences, University of Campinas (UNICAMP), Campinas, Brazil

P.F.D. Castro Department of Science and Technology Policy, Institute of Geosciences, University of Campinas (UNICAMP), Campinas, São Paulo, Brazil

S. Corder Department of Science and Technology Policy, Institute of Geosciences, University of Campinas (UNICAMP), São Paulo, Brazil

L.A.B. Cortez Faculty of Agricultural Engineering, UNICAMP, Campinas, São Paulo, Brazil; International Relations, UNICAMP, Campinas, São Paulo, Brazil

José Maria F.J. da Silveira Institute of Economics, State University of Campinas, Campinas, São Paulo, Brazil

M.E.S. Dal Poz Faculty of Applied Sciences, State University of Campinas, Limeira, São Paulo, Brazil

L.C. de Sousa Ministry of Development, Industry and Foreign Trade, Brazil

D.F. de Toledo Filho Ministry of Development, Industry and Foreign Trade, Brazil

A.F.P. Ferro Laboratory for Studies on the Organization of Research & Innovation, University of Campinas, Campinas, Brazil

W.M. Griffin Department of Engineering and Public Policy, Carnegie Mellon University, Pittsburgh, PA, United States

M. Harvey Centre for Economic Sociology and Innovation, Department of Sociology, University of Essex, Colchester, United Kingdom

I.R.L. Huamani Institute of Economics, State University of Campinas, Campinas, São Paulo, Brazil

P. Lemos Laboratory for Studies on the Organization of Research & Innovation, Department of Science and Technology Policy, University of Campinas, Campinas, Brazil

H. Lu 3E Information Development & Consultants, 3-eee.net, Wayland, MA, United States

H.L. MacLean Department of Chemical Engineering and Applied Chemistry, University of Toronto, Toronto, ON, Canada; Department of Civil Engineering, University of Toronto, Toronto, ON, Canada

F.C. Mesquita Department of Science and Technology Policy, University of Campinas, Campinas, Brazil

S. Salles-Filho Department of Science and Technology Policy, Institute of Geosciences, University of Campinas (UNICAMP), São Paulo, Brazil

I.T. Santos Center for Sustainability Studies of São Paulo School of Business Administration (GVces/EAESP), Getulio Vargas Foundation (FGV), São Paulo, Brazil

B.A. Saville Department of Chemical Engineering and Applied Chemistry, University of Toronto, Toronto, ON, Canada

S.C. Trindade SE^{2}T International. Ltd, Scarsdale, NY, United States

N.S. Vonortas CISTP and Department of Economics, The George Washington University, Washington, DC, United States; São Paulo Excellence Chair (SPEC) in technology and innovation policy, University of Campinas (UNICAMP), São Paulo, Brazil; ISSEK, Higher School of Economics (HSE), National Research University, Moscow, Russia

Introduction

S. Salles-Filho

This book discusses the unexpected path covered by the recent phenomenon called biofuels. Although exploring the plenty-of-opportunities world of renewables, the book proposes a realistic prospective analysis of biofuels, particularly of bioethanol.

It comes in a post-euphoria era over biofuels, just after the boom of investments in renewables and during a period of remarkable changes in the relative prices of energy and in the onset of the first formal agreement of the parties as regards to climate change.

In the following chapters the reader will find different and updated approaches discussing the futures of bioethanol. The optimistic forecasts made in the past decade have given way to more careful perspectives about the futures of renewables, particularly those based on agricultural feedstock such as corn, sugarcane and vegetable oils and animal fat.

There is a bit of contradiction in the biofuel recent trajectory. First, because they have been quite successful in many countries as in the cases of the United States, Brazil, and even the EU. In these countries/regions biofuels have performed very well mostly as blends, but also as direct fuels. Second, technical feasibility for first-generation biofuels is quite well developed and their economics if not highly stimulating, are, at least, fairly attractive. Third, biofuels have the advantage of tackling sustainability requirements in a fast and rather efficient way—yet are sometimes questionable.

The substitution of fossil fuels for biofuels, although far from the levels foreseen some years ago, is not negligible. In the United States, ethanol consumption represents about 8% of gasoline consumption. So, why are biofuels not evolving in a faster and generalized path and what are the trends and forecasts for the near future?

Throughout the chapters of this book that main issue is deployed in complementary questions. Will bioethanol become a global commodity? What are the expectations for second-generation bioethanol? Which roles will large companies play in biofuels and what trends can be drawn from their activities? To where are China, Germany, the United States and Brazil heading in this arena? What are the perspectives in Africa? How can the Paris Agreement on Climate Change transform the present trends? Which technologies are competing to define a technological trajectory in biofuels? These and other questions are addressed in this book by authors from different countries and with diverse approaches.

The book opens with the chapter of Wilson Araújo. This chapter presents a quite complete overview of the recent advances of bioethanol as a worldwide biofuel. Araújo's chapter is full of updated information and figures, sometimes focusing on technical aspects while other times deploying political and economic arguments. He emphasizes the importance of cellulosic ethanol as an essential driver of biofuels. Hundreds of millions of dollars have been invested in new technologies and in promoting second-generation bioethanol (E2G). However, national policies have been ambiguous, sometimes fostering, sometimes discouraging investments, and this places a big question mark on the forecasts.

The author is cautiously optimistic as regards to the trajectory of E2G. His main argument is based upon the cumulative investments and the learning curve already reached by companies and countries, and on the fact that E2G can really transform the scene in the short-to-medium term, and cope with the requirements of a reduction of greenhouse gas (GHG) emissions.

The following chapter is authored by Griffin, Saville and MacLean and focuses on the recent trajectory and perspectives of biofuels (highlighting bioethanol) with particular emphasis on the US case. The authors give a complete picture of the US developments in both bioethanol production and consumption, also presenting a wide and detailed overview of technical and economic potentials for bioethanol, other biofuels and biochemical products as well.

The chapter highlights the American regulatory framework, detailing the mandates towards biofuels and their technical and economic implications. They show how the mandates have provoked a shift in biofuel production and use in that country over the past decade. They also discuss the difficulties now faced by industries to cope with the mandate goals, particularly those related to cellulosic bioethanol. For them—as it is for the majority of the authors in this book— the perspectives are still open despite the many obstacles faced by bioethanol in recent years. Economic events (as for the fall of oil prices or the increase in new fossil sources) alongside technological advancements will shape the future of bioethanol and other biofuels as well. They are also cautiously optimistic, but as the American trajectory of the first generation has shown a vigorous performance—and unless extraordinary events arise—cellulosic ethanol is still a good bet.

Next, the reader goes into the European and Asian scenes, particularly from the perspective of land use change (LUC) and its consequences for biofuels. Harvey and Bharucha's chapter develops the trilemma approach in which the discussion of conflicts among food, energy production and climate change are addressed. Not only is this approach theoretically and empirically developed, but it is also analyzed looking at it in different situations. The authors discuss the trilemma issue in the EU—focusing on the German case—China, Brazil and India.

It is quite interesting to see how LUC matters in some countries but not in others. While in India and China this is clearly a matter of concern, in Germany it is probably less important and even less so in Brazil. In India biofuels are not one of the main goals depicted in its Intended Nationally Determined Contribution (INDC). Problems of pollution and food production are more likely to be about wind and solar than biofuels. The same can be said about China, but in this case shale oil and gas are the most probable competitors. Even in Europe where biodiesel is a reality, recent changes in the regulatory framework have provoked a reflux in the goals. In any case, Harvey and Bharucha show an improbable scenario of global diffusion of biofuels, at least when considering how countries are defining their priorities. Of course, their prospective is quite dependent on what countries will define as their contribution to reduce GHG emissions.

The chapter of de Sousa, Vonortas, Santos and de Toledo Filho proposes the view of "technological innovation systems" to analyze and compare the American and the Brazilian systems of innovation in bioethanol. This approach allows identifying how countries are evolving in the technological and market domain and how functional the systems can (or cannot) be. The authors show the differences between the systems, highlighting any expected disparities in terms of investment in R&D but mainly pointing to a mismatch among the components/actors of the Brazilian system when compared to the United States.

In this chapter one can see a more optimistic view of the futures of E2G. The authors argue that the present levels of technological development in both countries are reasonably close to each other and believe that both will probably go into the second generation. One important difference is that the United States is more likely to assume the technological leadership and sell E2G technology to Brazil.

The next chapter is authored by Salles-Filho, Bin, Drummond, Ferro and Corder. It presents an observed scenario of innovation inside the sugarcane–sugar–ethanol–electricity companies in Brazil. As also mentioned in other chapters, this sector is a traditional one in Brazil, historically composed of sugar mills and more recently by sugar–ethanol–electricity plants. In the beginning of the 2000s this sector experienced a boom of investments, and moreover, an important change in its capital structure, attracting foreign investors particularly from the oil and chemical companies. Shell, DuPont, BP, Dow among others, went on to take part in the "new" industry making substantial investments in the first and second generations. However, as is shown in other chapters, the last 5 years in Brazil have witnessed a reduction (if not a retreat) of investments both from national and multinational companies.

Despite the incoming investments in production and in R&D, innovation in this sector still is a secondary priority for companies. Most of them simply do not innovate at all, and even those that have innovation as a corporative priority—roughly speaking not more than 8% of them—make incremental movements guided by the prices of sugar, ethanol productivity and sugarcane productivity.

The chapter shows a mismatch between the subsystem of knowledge creation—public and private research centers, universities—and the industrial sector. Despite the quite impressive production of knowledge and new technologies (as to new varieties, for instance) indicators of productivity, cost reduction and portfolio diversification evolve at a very slow pace. Part of this is due to the lock-in entailed by the combination of sugarcane–sugar–ethanol–electricity. The predominant business model in Brazil allows companies to profit from different products creating pressure to innovate a dependent variable of a basket of prices, namely, sugar, ethanol and electricity. On the other hand, investments in E2G, although existing in pilot and even in industrial scales, still are pursuing technical and economic feasibility.

Luís Augusto Barbosa Cortez's chapter addresses precisely the present trajectory of the sugarcane industry in Brazil pleading for a change in its business model. The author argues that the present model in Brazil has achieved its limits and no huge expansion can be envisaged except the vegetative growth of internal market for ethanol and external market for sugar. Sustainability and diversification should be the main drive in this sector, and the use of cellulosic feedstock should be placed in the bull's-eye of the public policies and private strategies, he argues. Cortez presents convincing data and gives interesting alternatives toward a new business model to the ethanol–sugar–electricity industry in this country.

The Saville, Griffin and MacLean chapter amplifies Cortez's discussion, depicting a wide array of alternatives to explore cellulosic feedstock and hydrolysis technologies. The authors discuss technical and economic variables for using different feedstocks via diverse processes to obtain a collection of products and subproducts.

Feedstock as agricultural residues, woody biomass and dedicated energy crops are analyzed for their potentials. Analysis is also extended to process phases: pretreatment, enzyme hydrolysis and different alternatives of fermentation. Also the current situation of E2G development in the United States is presented alongside an analysis of the potentiality of cellulosic ethanol in that country. A comparison between technical features of ethanol from starch and cellulosic material is presented and reveals that there is a long path ahead to make E2G viable. The authors conclude that feasibility still depends upon improvements in alternative feedstock and in combining energy production with a basket of coproducts.

The next chapter presents a study of scientific and technological production using data and text mining tools. Authored by José Maria F.J. da Silveira, Luiz Gustavo Antonio de Souza, Maria Ester Dal Poz and Ivette Raymunda Huamani, the chapter shows an interesting scenario of knowledge production around the world related to second-generation bioethanol. The findings are quite interesting and reveal the concentration of scientific and technological leadership in the

United States and China. This comprises publications and patents, international collaboration and identifies countries and their leading institutions. Other countries such as Germany, Sweden, Finland and Brazil appear as important knowledge producers in this domain, although far from the two leaders. In Brazil, however, the connections of collaboration are fewer and weaker compared to other countries. Just one institution appears to be amongst the leading research organizations: the University of Sao Paulo. The chapter concludes with a question mark about the capacity of this country in assuming the technological leadership despite its sound scientific production.

Then, in the next chapter, Huaibin Lu presents the case of China and the perspectives of biofuels in that country. It shows an impressive stop-and-go movement in China, first with a clear national plan with mandates to blend ethanol to gasoline in some regions and then a retreat in the mandates. The major figures of China could imply huge demands for bioethanol and other biofuels, but as seen in Harvey and Bharucha's chapter, the trilemma problem, allied with the discovery of vast reserves of shale oil and gas, have shaped the conditions for setting different priorities. The perspectives of bioethanol depicted by the author are not very enthusiastic and perhaps China will turn its policies to energy efficiency, solar, and most of all shale gas and oil.

An overview on Africa, authored by Sergio C. Trindade, is the next chapter. He focuses on two countries: Kenya and Bénin. African bioethanol development seems to be some degrees behind the other countries analyzed in this book. Maybe one phrase in this chapter discloses the perception of the author: "the long-term future of biofuels in the world lies in Africa for its geographical location, resource endowment and increasing energy service necessities driven by development and population growth. But, in the short term, there has been limited market penetration of biofuels, ethanol and biodiesel in Africa's energy systems." Potential is the word.

Next, Paulo Lemos and Fernando Mesquita present a chapter with plenty of figures and forecasts that synthesize the present situation of production and consumption of bioethanol worldwide. The authors review data and forecasts concluding that "Bioethanol can be considered a restricted commodity because of its high importance in a few countries." Perspectives of becoming a global commodity do exist, but it is strongly dependent on a complexity of factors that cannot really be foreseen. "A second phase of commoditization is still far from being achieved," particularly considering the volume of international trade of bioethanol in the recent years and shown in this chapter.

Finally, Sergio Luiz Monteiro Salles-Filho writes a concluding chapter attempting to summarize the main arguments depicted in the book. In short, he argues there is a central line in the book leading the reader to conclude there is no definitive conclusion about the futures of bioethanol—and of other biofuels. On one hand, potentialities are still there, evoking an array of possibilities of producing different products and coproducts throughout different processes and targeting different markets. Ethanol can be obtained from diverse feedstock, two in particular are quite well known: starch from corn and sucrose from sug-arcane. On other hand, there are pressures coming from all sides: oil prices have dramatically fallen; shale oil and gas have suddenly become strong alternatives; LUC is a problem for some countries but not for others; the Paris Agreement came out to reduce GHG emissions; technology for E2G is already achieving industrial scale—or the same could be said otherwise: technology for E2G has not achieved technical and economic viability; economic feasibility for E2G depends on the generation of coproducts and use of new feedstock which is not yet well developed; first-generation biofuels can easily and widely be adopted and cope with goals of GHG emissions; electric engines for light vehicles is a reality; etc.

This complex situation, with several pros and cons, summarizes the discussions addressed in this book, placing the debate in a new position, not only based upon potentialities, but also carefully weighing the possibilities. Many scenarios can be built and this, we state, is the most recommendable way to analyze the futures of bioethanol: building and constantly reviewing alternative scenarios, as suggested by Sergio Luiz Monteiro Salles-Filho in the chapter of conclusions. The futures of bioethanol can indeed be many, with different trajectories living together. That is why we use the word future in the plural in this book.

Enjoy your reading.

Chapter 1

Ethanol Industry: Surpassing Uncertainties and Looking Forward

Wilson A. Araújo
DuPont Industrial Biosciences, Paulínia, Brazil

INTRODUCTION

The ethanol industry has been an important contributor to country economies and their energy security strategies besides playing a significant role regarding climate change issues. Its share in the global biobased market represents over USD 55 billion of a total biobased market of about USD 65 billion. Ethanol production is clearly an essential part of that industry which is no longer merely one for the future. Besides the well-stablished first-generation biofuels industry based on grains, beet and sugarcane (ethanol and biodiesel), investments are currently occurring around the world to produce second-generation biofuels (eg, cellulosic ethanol) and biochemicals (eg, farnesene, propenediol (PDO), etc.). Biotechnology advances and the large availability of biobased feedstocks, like agricultural residues, energy crops and even algae could boost biobased industry growth in addition to reducing greenhouse gas (GHG) emissions. Taking into account the global liquid fuel demand, biofuels could be considered an "easy think" business opportunity, but this is not a bed of roses. It is beyond simply planning and executing approach to attend a demand; this is not a business as "usual." It involves, for instance, scaling up technologies and establishing a learning curve towards economic feasibility, dealing with policymakers and oil industry lobby against biobased products as well as counting on global commitments towards a sustainable scenario under discussion in arenas pre and post the Paris Climate Change Conference (COP21).

The biofuels industry has reached the current stage thanks to parallel developments which boosted a surge in global demand for biofuels in the last decade: flex-fuel cars (ethanol and gasoline); policy mandates for greater use of biofuels; and subsidies in three most important regions: Brazil, the United States and Europe. The investment growth of biofuels is dependent on robust policies. Countries like the United States and Brazil have a long history of government support of their domestic industries and biofuel associations in both countries are very active in biofuels lobbying. It is important to note that policies have been globally fostering the expansion of the biofuels industry, but they are also often challenged to be reviewed, creating market uncertainties. Even in those two large biofuel producers the review of policies cannot be considered unusual and this debate is also often addressed in Europe.

Liquid biofuels account for the largest share of transport fuels derived from renewable energy sources. Ethanol production reached over 80 billion liters in 2013 and in 2015 was forecast to reach over 90 billion liters, with the United States and Brazil remaining as the top two producers. The demand growth and production have to evolve in a high entropy arena which is explored through this chapter beyond the first-generation biofuels domain. Based on market reports, academic studies and public information from industry associations and companies, this chapter pursues the understanding of cellulosic ethanol business dynamics as a viable alternative to increase global ethanol production. It provides a snapshot of the biobased economy and more granularly explores biofuels and ethanol industry growth challenges. First it presents an overview of the bioeconomy and market potential for renewable products. "Bioeconomy Toward a Renewable World" section brings the biofuels context, its global economic relevance and current industry reality. Finally, ethanol industry dynamics and the uphill task to grow and surpass policy and economic uncertainties are discussed. The challenge is to ensure the sustainability of first-generation business while accessing the second-generation ethanol potential.

Responsibility for the information and views set out in this chapter lies entirely with the author.

BIOECONOMY TOWARD A RENEWABLE WORLD

This section provides an introduction to the bioeconomy. It is a snapshot on world efforts toward renewables beyond the first-generation biofuels domain (biodiesel, sugarcane and corn ethanol).

Bioeconomy is a term which permeates academic and industrial audiences when discussing the challenge to build a sustainable future through biotechnology advances, the "modern" bioeconomy. In a broad definition, bioeconomy is the global industrial transition of sustainably utilizing renewable aquatic and terrestrial resources in energy, intermediate, and final products for economic, environmental, social and national security benefits (Golden and Handfield, 2014). Such a revolution requires coordinated strategies and commitments through legislation, policy, education and research. The United States and EU have been considering this theme as part of their strategic agenda. Brazil can be listed as a high-potential country due to comparative advantages, biodiversity, competitive costs of biomass—especially sugar cane and advanced tropical agriculture anchored in science and technology. Countries have been approaching the subject from different angles and defining its strategies, industry associations and international institutions have been coordinating efforts pursuing a deep understanding of growth hurdles and work on critical topics to drive success. It is a fair-minded approach to highlight that the world is struggling and, surely, progressing towards a renewable economy (Lorenz and Zinke, 2005; MEI/CNI, 2013; Pugatch Consilium, 2014; NAP, 2015).

The World Economic Forum has been working in convening cross-industry stakeholders at the regional level in North America, Brazil, India and China. This indicates both the growing importance of discussions about renewable energy and biobased consumer industry solutions around the world. However, despite recent advances and promises for the future, a significant amount of work remains, such as establishing effective government policy, increasing capital investment, influencing industry perception as well as connecting stakeholders across the value chain to reduce costs and stimulate growth and innovation. Country initiatives have also been noticed, for instance, in Brazil, the National Confederation of Industry (CNI) launched its agenda for stimulating innovation in Brazil in 2011. Among the strategic highlights are factors related to biotechnology and biodiversity. In the United States, the National Research Council supported by the Department of Energy convened an ad hoc committee to create a roadmap on industrial biotechnology for accelerating the manufacturing of biobased products (MEI/CNI, 2013; WEF, 2014; NAP, 2015).

The funding for this sector has been coming from governments (grants and loans), private equity and large companies. The total US federal spending on biofuels 2009–12 achieved USD 3.3 billion. Other national and state governments also fund the development of biorefineries, for instance, the National Development Bank of Brazil (BNDES) provided approximately USD 600 million between 2011 and 2014 for the financing of innovation in the Brazilian ethanol industry towards diversification (cellulosic ethanol, new products from sugarcane and gasification). CTC, Novozymes, Petrobras, Amyris and others have projects submitted and admitted in the BNDES program. Another global pot of funding comes from private equity investors like venture capitalists, angel investors and private individuals (NEXTSTEPS, 2014; BNDES, 2011, 2013).

The development of a biobased economy evolves in a complex context, climate change, production of biofuels, oil prices, population growth, environmental protection and the nexus with food security. Biorefineries will support this transition applying energy and cost-effective technologies to process biomass and produce renewables molecules. It involves a number of industrial sectors pursuing increases in productivity of agriculture and white biotechnology process advances. White, or industrial, biotechnology is the application of biotechnology for the processing and production of chemicals, materials and energy. Microorganisms, enzymes and their genetically engineered generations associated with process engineering strategies form the basis of a group of technologies that academy and industry are seeking to develop for commercial use. The biorefineries are combined biotechnological, chemical, physical and thermal processing facilities converting biological feedstock into numerous biochemical and/or chemical intermediates. For example, Amyris has successfully produced fragrance oil at its currently operating biorefinery in Brazil and shipped to its customer, Firmenich. Unlike the traditional sugarcane juice fermentation, which results in ethanol, they developed a biotechnology process (genetically modified microorganism fermentation) which applies the same juice to produce farnesene, a building block, which can lead to a range of sugarcane-based products like cosmetics, diesel, lubricants and others (Langeveld et al., 2010; Erickson et al., 2012; Elabora Consultoria, 2014; Youngs, 2014).

Biotechnology contribution to developing countries can achieve rates higher than the 2.7% of 2030 GDP estimated to OECD countries, with the largest economic contribution of biotechnology in industry (eg, high-energy-density biofuels produced from sugar cane and cellulosic sources of biomass) and in primary production (eg, genetically modified varieties) (OECD, 2009).

Numerous potential pathways to biofuels and biochemicals exist via the sugar platform. A consortium led by E4tech (UK) created a company database for 94 sugar-based products. It is a comprehensive evidence base for policymakers and industry—identifying the key benefits and development needs for the sugar platform. Some of the established biobased products already dominate global production (eg, ethanol, PDO, lactic acid), and several products do not have an identical fossil-based substitute (eg, xylitol, FDCA, farnesene). Twenty-seven products of particular interest were selected for further market analysis, given the level of industry activity, and as highlighted by the US Department of Energy "Top 10" biochemicals and International Energy Agency Bioenergy Task 42 reports. For those, the total current biobased market analysis achieved about USD 65 billion. The ethanol market is over USD 55 billion followed by much smaller, but still significant, markets for n-butanol (current production mainly via the ABE process), acetic acid and lactic acid, xylitol, sorbitol and furfural also showing significant markets for chemical conversion of sugars, without petrochemical alternatives. The smallest biobased markets are, as is to be expected, those of the earliest stage products, such as 3-HPA, acrylic acid, isoprene, adipic acid and 5-HMF. Biobased FDCA, levulinic acid and farnesene have the highest current prices (E4tech, 2015).

These products run the gamut from high-added-value products (low market volume) to low-value products (high market volumes, slim margins—biofuels). Scientific achievements have already boosted the execution of business strategies and brought products to market. Some cases in the United States illustrate those outputs, Cargill had a joint venture with Dow Chemical Company from 2000 to 2005 to produce polylactic acid, a biodegradable plastic made from sugar. In 2006, DuPont's joint venture with Tate & Lyle started up commercial production of bio-1,3-propenediol, a chemical used to make synthetic polymers, cosmetics, adhesives, detergent and antifreeze. In 2011, DuPont Tate &

Lyle Bioproducts signed a contract with Genomatica, a San Diego-based startup, to demonstrate fermentation of sugars from corn to bio-1,4-butanediol (BDO), USD 4 billion market potential for use in plastics, elastic fibers and solvent, BDO process was at commercial scale by 2013. The same year, BASF, announced results of a long-term investment in producing bioacrylate in partnership with Cargill and Novozymes, the prize was a piece of a USD 10 billion acrylate market (Elabora Consultoria, 2014; Youngs, 2014).

In 2009, a joint venture between BP and DuPont was formed, Butamax Advanced Biofuels. Butamax is developing isobutanol, which has a higher energy content that ethanol, can be blended at higher rates into fuel and also has the potential to catalyze the adoption of biofuels into the fuel supply chain at a faster rate, leveraging current infrastructure. Butamax's first commercial plant construction is in progress in Minnesota and the company strategy relies on the biofuel market. Gevo is another player in the isobutanol business, but considering the molecule more as a building block, not only for biofuel market. In 2014, Gevo started up isobutanol production, achieving around 400,000 liters/month in MN, USA. Blends of up to 16% of Gevo's renewable isobutanol were already commercialized for use in boats, motorcycles and snowmobiles (Carmann, 2011; Lane, 2014a; Gevo, 2014; Brandiwad, 2014).

Solazyme, a company producing oils from sugars with microalgae, has embraced the high-value and personal care market. Originally, funded by the Department of Energy and venture capital to produce advanced biodiesel, the company has raised cash with alternative bioproducts made from algal oil. Besides, their JV in Brazil with Bunge has manufactured the first commercially saleable products on full-scale production lines, oil and encapsulated lubricant (Youngs, 2014).

The high business profitability was probably a key driver in Solazyme's decision process towards the high-value market, but other elements can be addressed as part of such an equation and may also be observed in other company projects, for instance, typical challenges of scaling up of new biotechnological process, difficulties in keeping daily productivity and consequently limitations to supply large volumes required for the biofuel market. High-value products can strategically be considered as a first business phase to support the scale up of a new biotechnology process and the second phase could be a diversity of chemicals or concentrate the asset capability to produce a biofuel applying minor technological changes, for instance, BDO and PDO molecules previously mentioned illustrate pretty well such technological leverage. Although cautious, this approach to the execution in phases is clearly not a rule engraved in stone, for instance, Gevo and Butamax illustrate how companies approach markets in different manners. Financial results inevitably have to be achieved and, although company-specific, in some cases sharp strategic adjustments may have to occur to accelerate the delivery.

An analysis of the margin on a series of products would allow understanding the economic drivers that are most relevant to different companies. However, this is not usually released by corporations. In its absence gross margin can be calculation based on the annual statement of public companies. Recently, six companies were analyzed, Amgen, Roche (Pharmaceuticals) and Novartis have relatively high gross margins ranging from 66% to 82%, by contrast, Green Plans Inc. (biofuels) make only 6%. Solazyme and Novozymes achieved, respectively, 68% and 57% (NAP, 2015).

In developing innovative technologies such as second-generation biofuel production (eg, ethanol from sugar cane bagasse or corn stover) many businesses fail as they attempt to move from R&D and demonstration to commercialization. This difficult transition is often referred to as the "valley of death." Even though cellulosic and algal processes have been successfully demonstrated in pilot plants, few have been able to attract sufficient financing for commercial-scale facilities. Range Fuels began construction of a cellulosic ethanol plant in 2007 but shutdown in 2011 without having produced any fuel. In 2012, KL Energy Corp. produced a batch of approximately 75,000 L, a celebrated milestone. However, 1 year later, after changing its name to Blue Sugars, the company filed for bankruptcy. Other biofuel companies have lost significant value (30–90%) since they went public, for instance, Gevo, Amyris and Solazyme (ICCT, 2013). It is important to emphasize that projects with similar technological scope under capital-intensive companies like oil producers and chemical manufacturers are not immune.

The broad applications of advanced chemical manufacturing for multiple uses in energy, agriculture, food, cosmetics and environmental technologies can achieve trillions of dollars in addressable global market opportunities (NAP, 2015). Biofuels, the most noticeable outputs of the bioeconomy, are granularly tackled in the next sections towards ethanol analysis.

BIOFUELS

Biofuels have been intensively researched, produced and used over the past decade, in solid, liquid and gaseous states (Guo et al., 2015). Ethanol holds a significant role in the liquid biofuels market followed by biodiesel, so in this section an overview on liquid biofuels is presented.

Over the last decade there has been a massive public and private interest in diversifying energy sources, particularly after the sharp rise in oil prices in 2008. Triggers for this included volatility in oil prices, achieving a high level of energy security, reducing GHG emissions and promoting rural development. Brazil previously faced this effect in 1970s when energy supply became a political priority, which led to a robust federal ethanol program known as ProAlcool program, which is widely discussed in the literature. Domestic markets have been the pillar to successfully implement biofuel strategies with international trade emerging from them afterwards. Biofuel internationalization started in the 2000s and currently ethanol and biodiesel are established products, traded daily and globally. Three parallel developments boosted a surge in global demand for biofuels in the last decade: (1) the advent of the flex-fuel cars (for ethanol and gasoline); (2) new policy mandates pushing for greater use of biofuels as alternatives to fossil fuels; and (3) subsidies in the three most important regions: Brazil, the United States and Europe (Goldemberg and Nogueira, 2014; Martins and Gay, 2014; UNCTAD, 2014).

Liquid biofuels (primarily ethanol and biodiesel) account for the largest share of transport fuels derived from renewable energy sources. The share of renewables in transportation remains small. Renewable energy accounted for an estimated 3.5% of global energy demand for road transport in 2013, up from 2% in 2007. Biofuel's contribution to the transport sector is considerably higher in some European countries, in the United States, and Brazil—where the share of biofuels in road transport fuel exceeded 20% in 2014 (REN21, 2015).

Biofuels have been driven by governmental policies. The United States and the European Union have some of the world's most aggressive policies for alternative fuel promotion, including volumetric mandates. The key instruments widely adopted to foster production and consumption have been mandatory blending targets, tax exemptions and subsidies. Governments have also intervened on the production chain by supporting intermediate inputs (feedstock crops), subsidizing value-adding variables (labor, capital and land) or granting incentives that target end products. In the United States, the world's largest ethanol producer, strong financial incentives are guaranteed for biofuel manufacturers. In the European Union, the world's largest biodiesel producer, biofuel consumption is mostly driven by blending mandates. Policymakers and industry executives are constantly interacting in that arena, pursuing the environmental compliance and industry growth. In this chapter policies are not addressed in depth, but the theme will permeate the discussion on ethanol market dynamics (Sorda et al., 2010).

In 2014, USD 270.2 billion were globally invested in renewable energy initiatives. It is nearly 17% higher than the previous year, excluding large hydroelectric projects. Biofuels presented an 8% fall in global investment to USD 5.1 billion contributing just fewer than 2% of overall global investment in renewable power and fuels. This survey included all biofuel projects with a capacity of 1 million liters or more per year. It was expected that oil prices would have impacts in 2015 investments in a few places, such as developing countries burning oil for power and biofuel markets not covered by mandates (UNEP/BNEF, 2015).

Renewable energy employed 7.7 million people, direct or indirectly, around the world in 2014. Liquid biofuels remain a large employer, accounting for nearly 1.8 million jobs worldwide. Brazil has the largest workforce, with 845,000 employed. Job losses in the ethanol industry (due to the increasing mechanization of sugar cane harvesting) were more than offset by job growth in biodiesel, mainly supported by incentives such as increased blending requirements (IRENA, 2015). Other major biofuel job markets in Latin America include Colombia and Argentina, with workforces of 97,600 and 30,000, respectively.

The United States, France and Germany are key biofuel producers, though mechanized harvest and processing limit employment compared to countries with more labor-intensive operations (IRENA, 2015).

Global biofuel production in 2013 reached over 115 billion liters. A challenging path is expected for transport biofuels to 2025. Overall biofuel production must triple, and advanced biofuels need to increase 22-fold to meet the goal targets of limiting the global increase in temperature to 2°C by limiting the concentration of GHGs in the atmosphere to around 450 parts per million of CO_2 (IEA, 2014). Ethanol and biodiesel production are both expected to expand to reach, respectively, 158 billion liters and 40 billion liters by 2023 (OECD-FAO, 2014). Ethanol and biodiesel will continue to be mostly produced from feedstocks that can also be used for food. By 2023, 12%, 28% and 14% of world coarse grains, sugar cane and vegetable oil production, respectively, are expected to produce first-generation biofuels. Those figures help to understand the issue as production is not even expected to double by 2023 (115 vs 198 billion liters). Roughly, 100 billion liters of production gap can be foreseen. The European Union is expected to be by far the major producer and user of biodiesel. Other significant players are Argentina, the United States and Brazil, as well as Thailand and Indonesia. The three major ethanol producers are expected to remain the United States, Brazil and the European Union. The biggest well is in an early stage of exploration—cellulosic material, the woody part of the plant and potentially abundant source of energy. The race is in optimizing the technology that can produce biofuels from cellulosic sources more efficiently; biotechnology companies are in a race to deliver the second-generation biofuels. In Brazil and the United States, companies are using sugar cane biomass (bagasse and straw) and corn stover as inputs to run their first commercial second-generation ethanol plants, policies have been critical to foster the industry growth. Biofuels made from nonfood sources such as agricultural, municipal, and forest waste, high-yielding cellulosic crops and algae are potentially important low-carbon liquid fuel options (Shubert, 2006; EurObserver, 2014; IEA, 2014; OECD-FAO, 2014).

Biofuels will continue to be part of a global agenda. A global climate agreement was recently and finally achieved at COP21, the Paris Climate Change Conference. The Low Carbon Technology Partnerships initiative (LCTPi) was one of the important efforts prior to COP21 launched pursuing a series of concrete action plans for the large-scale development and deployment of low-carbon technologies. Although the increase in electric cars using renewable energy will help reduce emissions from light vehicle transportation, liquid fuels will be the most likely main alternative for personal transportation for a long time to come. Hydrocarbon fuels will be the most likely choice for aviation, maritime and trucking for the main part of this century. Given these conditions, it is essential that we seek to replace fossil fuels with advanced biofuels that reduce CO_2 emissions by at least 50% (eg, cellulosic ethanol and sugarcane ethanol). Development of a new advanced biofuels market will speed up emissions reduction, recruit investment into the sector and drive the growth of high-tech green jobs across the globe (LCTPi, 2015).

ETHANOL INDUSTRY: THE UPHILL TASK TOWARD A SECOND-GENERATION SCENE

Ethanol is liquid fuel typically made from biomass, sugarcane, cereal grains and sugar beet. It can replace gasoline in modest percentages for use in ordinary spark-ignition engines (stationary or in vehicles), or can be used at higher blend levels (usually up to 85% ethanol, or 100% in Brazil) in slightly modified engines such as those provided in "flex-fuel vehicles." Note that some ethanol production is used for industrial, chemical and beverage applications and not for fuel—this chapter is fuel-oriented.

The ethanol production reached over 80 billion liters in 2013 having as top-five producers the United States, Brazil, China, Canada and France. Europe remains a relatively modest player globally at about 6 billion liters production. The United States, Brazil and Europe have been using, respectively, the following feedstocks: (1) corn; (2) sugarcane and corn as a recent option; and (3) wheat maize and sugar beet (BNDES, 2014; ePure, 2014; REN21, 2015). World ethanol production is presented Fig. 1.1 (ISO, 2014).

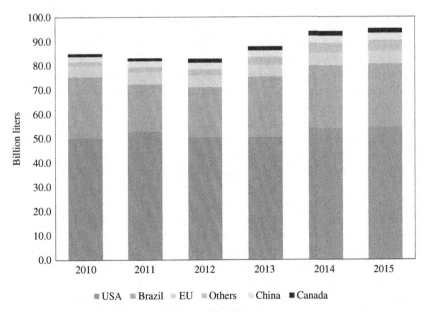

FIGURE 1.1   World fuel ethanol production (ISO, 2014).

In 2014, the United States exported 2.9 billion liters, overtaking Brazil, as the world's largest exporter, with 980 million liters. Most of the markets barely existed 8 years ago (eg, Canada, Philippines, South Korea, etc.). Strong industry economics, competitive and abundant feedstock, and US biofuels mandate uncertainty (blend wall) made its exports attractive, for both the US industry and customers. The export market shares of those two main producers are: Brazil (19.5%) and the United States (45.2%). The export forecasts for 2015 to United States and Brazil were, respectively, 2.8 billion liters and 700 million liters. Potential new markets focus is in Asia, as ethanol usage is low and fuel consumption growth is the fastest in the world (eg, China, India, Philippines, etc.) (Dwyer, 2015). Regarding import tariffs, the European Union decided to impose taxes on US ethanol imports for a 5-year period pushing a number of American producers to find a way around having to pay these duties by routing their production via Norway, which in turn exported the biofuel to the European Union in the form of a gasoline–ethanol blend. ePure alerted the European Commission to this practice, and in 2014 the Commission decided to apply the antidumping duties on all American ethanol regardless of transit country. Brazilian exporters adopted a similar strategy, shipping hydrous ethanol to Central America and Caribe to be dehydrated and shipped to the United States avoiding duties, the USD 0.54/gallon tariff on imported ethanol expired in 2011 (UNICA, 2011; EurObserver, 2014).

The world fleet of 800 million light-duty vehicles (LDV) in 2010 could reach 2 billion cars in 2050. The United States and Brazil concentrate the ethanol fuel consumption. Blends of 10–20% ethanol in gasoline have proven technically feasibility in many countries. However, globally 75% of ethanol consumption in transportation is usually limited to 10%, named E10. In Brazil, flex-fuel vehicles represents over 60% of LDV and 20% of motorcycle fleets. Ethanol (hydrous E100 and anhydrous) represents approximately 40% of the country's liquid fuel consumption (Otto cycle vehicles). E100 can be used nowadays by 23.8 million Brazilian vehicles (mostly cars with flex-fuel engines). The US fleet included 14 million flex-fuel vehicles, of which more than 10% were using E85 (blends containing 51–83% ethanol, as lower levels are used during winter months to ensure cold starting). EU gasoline typically contains up to 5% ethanol and is widely available as the default choice. E10 is available in France, Finland and Germany and other European countries are considering its introduction (UNICA, 2014; IEA, 2014; Martins and Gay, 2014; Souza et al., 2015).

The Brazilian sugarcane ethanol industry directly employs 613,000 people and when seasonal positions are taken into account that number achieves 988,000 employed. Corn ethanol industry in the United States employs 232,000 of which around 90,000 are directs (Markestrat, 2014; Urbanchuk, 2015). The European ethanol industry has created and sustained 70,000 direct and indirect jobs. With unemployment—particularly in youth and rural unemployment—remaining a key concern in Europe, the ethanol industry serves as a vital ingredient in Europe's quest to boost growth and create high-skilled and "green" jobs. Ethanol companies have made investments in Europe over recent years totaling €8 billion (ePure, 2014).

The Renewable Energy Directive (RED) and Fuel Quality Directive (FQD) are the two main policies driving biofuel deployment in the EU out to 2020. The last few years have seen a plateau in the EU's consumption of biofuels, and minimal deployment of advanced biofuel routes, primarily due to policy uncertainty surrounding indirect land use change. The EU set ambitious objectives of achieving 10% renewable energy in the EU transport sector and a 6% reduction in the GHG intensity of fuels used in road transport. In order to qualify for both of these targets, biofuels consumed in the EU must comply (and demonstrate compliance) with strict sustainability criteria (ePure, 2014; Mousdale, 2014). International current events involving conflicts in the Middle East and the Ukraine–Russia crisis could prompt the EU to adopt a more proactive policy to its reliance on hydrocarbons. According to the European Commission, the EU imports 94% of its oil consumption and 30% of these imports of crude oil and refined products are sourced from Russia (EurObserver, 2014).

The United States and Brazil have a long history of government support to their domestic industries. Both the American and Brazilian sectors have been operational since the 1970s. Although the United States has been the world's largest producer of ethanol since 2005, it protected the domestic market from Brazilian competition until 2011. The divergent interests of the key stakeholders engaged in the biofuels dynamics require a measure of reconciliation assuming biofuels are to be produced and used at a national level. The stakeholders are the producers and traders, oil refiners, farmers, automakers and dealers, oil marketers and retailers, governments and consumers. It is important to note that industry associations of Brazil (UNICA) and the United States (RFA) have been very active in biofuels lobbying. In Europe, ethanol is the most taxed fuel in the EU's entire energy mix, and a fairer taxation system based on energy content and CO_2 performance has been advocated by industry (ePure) to support growth of the sector. Although policies have been globally fostering the expansion of the biofuels industry, they are also often challenged to be reviewed creating market uncertainties. A robust market for first-generation biofuels, combined with a long-term investment plan for advanced biofuels, is needed to release the global bioeconomy potential (ePure, 2014; Farina, 2014; Sorda et al., 2010).

The absence of a Brazilian long-term policy to ethanol corroborated to the United States surpassing its main competitor. The main Brazilian government misconceptions, gasoline subsidized prices to control inflation and past reduction on ethanol blending (19 months, 25–20%, Oct. 2011 to Apr. 2013), severely impacted the sugarcane ethanol sector. This was also echoed in the global sugar market. Roughly, 3 billion liters were not blended in gasoline, equivalent to 5 million metric tonnes of sugar which was delivered at the international market, contributing to the high global inventory and consequently a decrease in sugar prices. This sugar production alternative was possible because sugarcane is used to produce ethanol and/or sugar, which is beneficial for the producer, who can count on alternatives in the event of demand/supply shocks for the product, plant configurations usually allows sets 40:60 or 60:40 to manage feedstock destination. There are also sugarcane-based plants configured to produce only ethanol (Farina et al., 2010; Nastari, 2015a,b; Serigati and Possamai, 2015).

The Brazilian industry is experiencing a moment of relief, but the argument for a long-term policy is still valid. A long-awaited increase in Brazilian ethanol blend rate with gasoline from 25% to 27% became effective this year. This measure is expected to increase domestic consumption of ethanol by 1.2 billion liters. In addition, federal taxes were increased on gasoline (R$ 22 cents/L~USD 7 cents/L) and diesel (R$ 15 cents/L~USD 5 cents/L). As a consequence, ethanol prices which are closely linked to fossil fuel prices at the retail pump will potentially provide revenue for the financially strapped ethanol sector. Tax and mix strategies are positives, but with limited impact, only more efficient business groups and in a certain indebt level will really leverage this momentum. The producer's survival challenges are on recouping the financial sustainability of sugar and ethanol in the short term boosted by credible government measures. Sustainable returns will provide the robust conditions to the sugarcane ethanol industry struggles and recoup the track of investments. It is important to note that even in a turbulent scenario UNICA reported that the sector invested USD 2 billion in mechanized harvesting and sugarcane field renewal since 2006. Although impacts of policies in the Brazilian scenario are described here, this cannot be considered a Brazilian specificity, stable and predictable policies are very important for the global ethanol industry (Farina, 2015; USDA, 2015).

World ethanol prices are projected to increase by 9% in real terms from 2014 to 2023 (OECD-FAO, 2014). Two elements were taken into account as the main influencers in the level of ethanol prices. First, an increase in market-driven demand for hydrous ethanol (E100) by owners of flex-fuel cars in Brazil is expected, given the assumption of strong crude oil prices and of Petrobras not freezing the retail price of gasoline anymore (government-influenced). Second, policies in place such as the 25% blending requirement in Brazil and the level of advanced gap in United States should also reinforce ethanol prices. The domestic US corn-based ethanol price should not increase as much as the Brazilian world ethanol price and 8% were expected to be exported in 2013. It is already, partially, possible to crosscheck projections, reached exportation and projected were, respectively, 4% and 8%. Regarding Brazilian market expectations, besides recent 27% blend and tax increases on gasoline, hydrous ethanol (E100) demand can also help to boost Brazilian sugarcane sector gains and this growth may also interfere in sugar prices contributing to overall business sustainability (OECD-FAO, 2014; Neves, 2015).

In the United States the maximum amount of ethanol that can be mixed with gasoline in low blends is 15% for cars built after 2001. Since older cars will eventually leave the fleet, the amount of ethanol being consumed in low-blend mixes is forecast to increase over the next decade to reach a maximum level of 14% by 2020. However, this assumption is subject to uncertainty as at present the supply of E15 blends to consumers is encountering some difficulties. There are different reasons for this: retailers may not be willing to supply E15 due to the fact that earlier car warrantees may limit ethanol content to a previous 10% limit, misfueling of vehicles by consumers or simply problems of availability at the pump. The quantity of ethanol to be used over the next decade in the United States will be limited by the blend wall and by the expected decrease in gasoline consumption (OECD-FAO, 2014). Less than 0.5% of ethanol is consumed in the form of E85. The near-term potential market for E85, if attractively priced, is over 45 billion gallons (170 billion liters) (Light, 2014). In Dec. 2007, US Congress passed this major new energy efficiency and new, much larger RFS, now known as Renewable Fuel Standard 2 (RFS2). The RFS, except for biodiesel, is expressed in gallons of conventional fuel (mainly corn ethanol). The mandate comprises four nonexclusive submandates that are defined by feedstock and lifecycle GHG savings compared with petroleum: (1) renewable fuels (at least 20% GHG savings); (2) advanced biofuels (at least 50% GHG savings); (3) biomass-based diesel (at least 50% GHG savings); and (4) cellulosic biofuel (at least 60% GHG savings). Mandate volumes for all categories other than cellulosic fuel have been met every year. Cellulosic volumes have been revised downward in each year due to low availability. The recent RFS volumes included substantial reductions from the statutory standards in the original 2007 version, volumes set to 2016 are: cellulosic biofuel—0.206 billion gallons (0.78 billion liters); advanced biofuel—3.4 billion gallons (12.9 billion liters); total renewable—17.4 billion gallons (65.9 billion liters); implied corn ethanol—13.8 billion gallons (52.2 billion liters). The gasoline consumption expected to 2016 (EIA projections) is about 377 million gallons a day (1.4 billion liters), 138 billion gallons annually (522 billion liters). Thus, if it was possible to blend 10% ethanol everywhere and in all seasons, the effective blend limit would be 13.8 billion gallons (52.2 billion liters), (Tyner, 2012; Mousdale, 2014; Lane, 2015a,b; EPA, 2015).

Compliance with the RFS is measured using Renewable Identification Numbers (RINs). When qualifying biofuels are produced, each gallon is assigned a RIN. Until the biofuels are sold as fuel or blended into conventional fuels, the RINs are "attached" to the fuel. Once the biofuel has been blended or sold, the RINs are detached, and can then be bought and sold like other commodities. At the end of each year, fuel suppliers must calculate their renewable volume obligations based on their total gasoline and diesel sales, which indicate the total number of each type of RIN that the suppliers must submit to EPA. To the extent that a supplier has excess RINs, that supplier may sell them to others who may be short, or save them for use in the following year. The RFS2 policy intervention changes the natural market dynamics. It mandates that a certain quantity of biofuels be used; in other words, it aims to raise demand beyond the equilibrium level supported by market. At this quantity, there is a gap between the price at which suppliers can afford to sell the fuel and what the consumer is willing to pay for it. To close this price gap there is RIN mechanism which adds value to the fuel for the producer to close this price gap. When RFS2 was created, it was expected that biofuels would be more expensive to produce than fossil fuels, and thus the targeted volumes of biofuels would not be blended for market reasons alone. The RFS2 adds value through RINs to make biofuels cost-competitive. It was also expected that cellulosic fuels would be even more expensive than other renewable fuels and thus would require a stronger policy signal to bring them to market in desired volumes. RINs are differentiated by biofuel type in the same way as the main categories RFS2 mandated volumes. Due to the low availability of cellulosic ethanol, EPA also sets waiver credit prices for cellulosic ethanol in order to allow obligated parties to meet their required volumes. In addition to RINs, when the supply of cellulosic fuel is expected to fall short of the volume mandated by the RFS2 in any given year, the EPA is required by statute to reduce the volume of cellulosic biofuel required for compliance. This is referred to as waiving part of the mandate. In this period, EPA is also required to make cellulosic waiver credits (CWCs) available to obligated parts as an alternate compliance option for any number of cellulosic RINs up to the revised volume mandate. CWC prices were USD 0.49 and USD 0.64/gallon in 2014 and 2015, respectively. Basically, there is a policy to ensure a degree of confidence to producer that its biofuel will be sold competitively. This price expectation should serve to lower the risk of investment in cellulosic ethanol nascent industry. It has imposed a system of credits that gasoline refiners and importers must purchase for each gallon of mandate cellulosic ethanol they fail to blend. In Brazil, BNDES has been working with industry to build the cellulosic ethanol momentum by evaluating policy alternatives to be discussed with government. The bank has been consistent, even with no specific Brazilian policies to cellulosic ethanol, loans have been approved to commercial-scale projects in Brazil: Raizen, USD 91 million GranBio, USD 150 million; and Abengoa, USD 116.8 million. Abengoa also received a USD 132.4 million loan guarantee and a USD 97 million grant through the Department of Energy to support construction of their facility in the United States (Law360, 2010; Yacobuci, 2013; ICCT, 2013; Lane, 2014b; Standlee, 2014).

First-generation ethanol produced from sugars or starch directly extracted from biomass has limitations. For example, corn ethanol comes with problems, it offers only modest savings in GHG emissions compared to gasoline (Peplow, 2014). Significant concerns have been voiced about the impact of large-scale production of first-generation feedstocks on food availability, food and animal-feed prices, deforestation and water resources, as well as about the net impacts of some biofuel pathways on climate change and air pollution. A next-generation ethanol economy is on its way, the second-generation based on nonfood feedstocks. Innovations related to advanced feedstock options, including dedicated energy crops, waste sources and algae are emerging in the market. Although much attention has been focused on conversion technology, biochemical and thermochemical processes, feedstock is equally critical as quality and access to feedstock supply is the first bottleneck to overcome commercial production (Huenteler et al., 2014; Service, 2014; Zhang, 2014).

Global projections for cellulosic ethanol revealed a huge potential. Based on transport fuel demands and agriculture residue availability, two scenarios were evaluated, what it will take to replace 10% of gasoline demand with cellulosic and how much gasoline could be replaced with cellulosic ethanol if the available agricultural residues (17.5%) were all converted into fuel. Results were, respectively, 115 and 351 billion liters/year. Five major crops represent about 90% of the 24 crops analyzed, totaling 800 million dry tonnes in 2030 available for cellulosic ethanol conversion (wheat, maize, sugarcane, rice, soybeans). Based on the major residue potential the top four projected availabilities are located, respectively, in China, the United States, Brazil and EU-27. The study outputs related to investment and installed capacities are presented in Table 1.1. The following risks across the cellulosic ethanol value chain were listed: feedstock supply risk, fragmented supply chain, insufficient infrastructure, high capital costs, technology risks, product delivery risk and market access limited. The main results for industry challenges are presented in Fig. 1.2 (BNEF, 2012). All topics listed in Fig. 1.2 allow realizing the main areas to be addressed, it is a nascent industry and there are no shortcuts to surpass such complexity toward substantial production volumes. The overall potential of this value chain can be achieved only through a heated, but convergent, synergy of industry stakeholders. Industry efforts on improving business performance (technology, feedstock logistics, learning-by-doing, etc.) linked to temporary incentives and long-term climate change aligned policies are important factors to minimize market uncertainty and so fuel that growth process.

TABLE 1.1 2030 Cellulosic Ethanol Potential

Category	Brazil	China	USA	EU-27
Biomass availability (million metric tonnes)	177	221	180	151
Fuel demand (billion liters)	8	24	53	9
Residue potential (billion liters)	71	89	72	60
Investment-installed capacity (USD billion): fuel demand scenario (residue potential)	11 (94)	32 (118)	71 (96)	12 (80)
Revenue throughout 20-year plant lifetime (USD billion), at USD 0.44/liter	75 (622)	213 (779)	469 (633)	78 (532)

Source: Adapted from BNEF, 2012. Moving Towards a Next-Generation Ethanol Economy. Available from: https://www.dsm.com/content/dam/dsm/cworld/en_US/documents/bloomberg-next-generation-ethanol-economy.pdf.

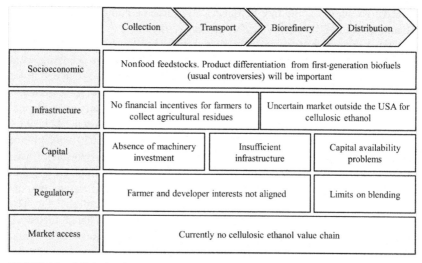

FIGURE 1.2 Cellulosic ethanol industry challenges (BNEF, 2012).

Cellulosic ethanol is particularly promising because it can capitalize on the biotechnology advances to severely reduce costs, is derived from low-cost and plentiful feedstocks, can achieve high yields vital to success, and it is environmentally friendly. A pretreatment opens up the biomass to enzymes that breakdown the hemicellulose and cellulose, which comprises ~20–30% and ~40–50%, respectively, of the material, into sugars that are fermented into ethanol. Lignin and other components not converted into useful products can be burned to provide heat and electricity to run the process, with the excess sold. This technological route is known as sugar platform and there are also thermochemical and carboxylate platforms (Wyman, 2007; Brown and Capareba, 2015; Wyman and Dale, 2015; Holtzapple et al., 2015). The thermochemical platform, a combination of temperature, pressure and chemistry, can produce either a crude oil or a stream of carbon monoxide and hydrogen known as a syngas. After further treatment and refining with the help of catalysts, both can turn into hydrocarbons such as gasoline, diesel and jet fuel. These "drop-in fuels," named because they can replace normal fuels with no adjustments to engines, have been claimed by some to be a more appropriate goal to pursue mainly because in this case they have no blend wall to vault (Krieger, 2014; Peplow, 2014).

The cellulosic ethanol production costs have often been put on the spot. However, although thousands of uneconomic approaches can be conceived, only a team with demonstrated experience in the design, construction and operation of commercial biological and biomass-based technologies is likely to design state-of-the-art, commercializable and economical processes. To further complicate matters, many of the best unit operations involve knowhow and trade secrets (Wyman and Dale, 2015). For example, most recently BNDES conducted a study on cellulosic ethanol feasibility and potential policies to foster this industry in Brazil (BNDES, 2015). Industry players like DuPont, DSM, Novozymes, Beta Renewables, CTC, Abengoa and others attended BNDES call to discuss technology yields, capital and cost of manufacturing figures, those companies contributed with limited insight during meetings, the trade secret obstacles pushed study coordinators to build plant cases supported with state-of-the-art process simulations tools and limited corporate inputs. Thus, it is relevant to state published estimates on costs and capital do not accurately capture strategic corporate numbers, the ones critical to business success. At least the discussions throughout the study elaboration allowed BNDES to fine-tune assumptions and define cases, valuable interactions with players, which of course has strong interest to globally develop such industry. Cost projections publically and globally available can provide valuable benchmarks with which to target technology development; no process cost is ultimately meaningful until it is validated by commercial success, projections must be viewed as just that.

For many products and services, unit costs decrease with increasing experience. First-generation corn and cane ethanol have declined considerably over recent decades. In both the United States and Brazil ethanol costs have declined by approximately 60%. Simulations assuming policies will be in place to foster cellulosic ethanol achieved cost reductions of about 30–70% between 2015 and 2035 (Goldemberg et al., 2004; Hettinga et al., 2009; Bake et al., 2009; Xiaoguang et al., 2012).

Projected capital costs for the first cellulosic ethanol ventures are high. Those numbers are influenced by overdesign of initial projects to mitigate lack of large-scale experience with technology. Government policies promoting first-of-a-kind applications are needed to overcome these major impediments for current technologies, just as the petrochemical industry grew, out of necessity, through government support during World War II. Recent analysis has concluded that the USA federal commitment to oil and gas was five times greater than the federal commitment to renewables during the first 15 years of each subsidies' life, and it was more than 10 times greater for nuclear, the historical average of annual energy subsidies, when looked at through the lens of more than a century of federal support for energy, shows a similar result. It is not just an historical view, current there are eight tax breaks for oil companies with an annual cost to the USA government of more than USD 4 billion. It is not realistic to state that current renewable energy subsidies constitute an oversubsidized outlier (Wyman, 2007; AEC, 2015).

The cost matter has been widely evaluated by governments, investors and academics. At the NREL findings on economics, in total, operational expenses, and allowing for profit and returns on total capital investment, has been estimated to be about USD 2.15/gallon (USD 0.57/liter). High yields are critical to lower costs and are a necessary, although not sufficient, requirement. Feedstock, enzyme and capital costs were listed as part of the challengeable equation. Current enzyme costs are up to USD 10.00/kg of enzyme protein, enzyme can translates into a cost about USD 1.00–1.50/gallon ethanol (USD 0.26–0.40/liter). In a feedstock example where cost would be about USD 80.00/ dry ton cost, feedstock impact could achieve about USD 1.00/gallon (USD 0.26/ liter). Capital investment projections are about USD 7.00/annual ethanol gallon (USD 1.85/liter), including on-site enzyme. Amortizing these costs would contribute approximately USD 1.00/gallon (USD 0.26/liter) (BETO, 2014; Wyman and Dale, 2015). Some corporate numbers are available to capital expenditures ranging in dollars per annual gallon from USD 7.00–10.00 (USD 1.85–2.64/ liter) (NEXTSTEPS, 2014).

In Brazil, BNDES economically evaluated some plant configurations. Among those cases a promising mid-term case 2021–25 was evaluated: energy cane as feedstock at USD 40.00/dry metric tonne (high cellulosic material content, not typical sugarcane varieties), 284 L ethanol/dry metric tonne (75 gallons/dry metric tonne), electricity production 210 kWh/dry metric tonne, production of 411 million liters/year (109 million gallons/year), R$ 922 million capital (USD 400 million) equivalent to about USD 1.06/annual ethanol liter (USD 4.00/gallon), enzyme cost impact USD 0.035/liter of ethanol (USD 0.13/ gallon), overall cost of ethanol about USD 0.30/liter equivalent to USD 1.20/ gallon (labor, consumables, biomass, enzyme, capital and maintenance costs) (BNDES, 2015).

A short-term view 2016–20 example, smaller capacity, and as feedstock only bagasse surplus from a small typical sugarcane ethanol plant, presented an overall ethanol cost of USD 0.63/liter of ethanol (USD 2.38/gallon of ethanol) and annual production of 92 million liters (24.3 million gallons). The short-, mid- or long-term approaches are study-specific, BNDES adopted its own to categorize the 14 cases evaluated in that work (BNDES, 2015).

All those technoeconomic figures are just examples to illustrate cost dynamics, it is important to emphasize cost estimates are very specific to the business case and technology performance is a critical variable in it. The promising BNDES mid-term example associated with comments captured during interaction with industry representatives may help to speculate about current cellulosic momentum in Brazil. For some technology providers scenarios were considered conservative once they claimed to be able to achieve yields over 300 ethanol liters/dry metric tonne of sugarcane bagasse. In other words, a few players are confident they have an economically sustainable technology, but the operational experience challenge remains, to be gained throughout the coming decades running the first wave of plants and accessing the learning-by-doing cost-benefits.

Production costs of alternative transportation fuels have also been evaluated by International Agency Energy (IEA), the production costs of 20 fuels were examined for crude oil prices between USD 60 and USD 150/barrel of oil. USD 60/barrel was the reference point as a long-term series of data were linked to low oil prices. The study was based on a set of technical and economic parameters such as conversion efficiency, energy density of feedstocks, scaling factors, lifetime of infrastructure and conversion facilities, interest rate, historical cost movements and others. The study results showed that many alternative fuels can compete at USD 100/barrel in mature technology scenario—technologies fully benefited from economies of scale or know-how. Second-generation fuels can also achieve cost-competitiveness with gasoline, corn stover as feedstock and biochemical conversion were considered to cellulosic ethanol case. Feedstock prices play a major role in the final (untaxed) costs of alternative fuels. Corn ethanol is not as cost-competitive with petroleum fuels when the oil price is close to USD 60/barrel. With oil prices close to USD 150/barrel, its cost-competitiveness depends on corn prices. Sugarcane ethanol remains cost-competitive with petroleum fuels at USD 60/barrel (IEA, 2013). The production costs are definitely relevant to business development as well as the strategic view on role of biofuels (eg, ethanol) taking into account the trade-off between investment risk today and carbon emissions reductions in the future, stakeholders are forced to adopt a holistic view where policymakers play a significant role toward balance. During the chapter elaboration the price of a barrel of oil was around USD 50.

Although many advanced biofuel and biochemical companies have lost value in the market since their IPOs, such devaluation, by itself, is not a fair metric of the success of these firms. It is not uncommon for companies to lose significant value if they go public before stabilizing profits. Those companies launched their projects only some years ago; some are still fine-tuning their process equipment, while others may still be building out production capacity. For example, Solazyme and Amyris only recently made substantial progress from a steady-state production perspective. Macroeconomic trends considerably impacted those advanced biofuel companies, such as slow recovery from financial crisis, restrained investment due to recessions, and the European sovereign debt crisis in 2010–11. External factors may also have an influence, such as oil and corn prices, changes in government funding, changes in the volumetric mandates and other relevant legal instruments and public interest in those biofuels (ICCT, 2013).

The vast majority of funds from government grants/loans, private equity, and large Fortune 500 companies, have been dedicated to projects oriented to major technological breakthroughs in cellulosic and algae-based pathways at new, standalone biorefineries. Large companies typically have longer planning horizons than small, venture-funded startups. They have the advantage of continuing funding a project even when there are problems or delays. At present, there is no durable market signal to make the case for a shift of investment into biofuels by these companies. Their profit margins are simply too large in their core businesses. A glimpse at investments allows to realize the comings and goings: recently BP left the cellulosic ethanol arena and turned its focus to Brazil, pursuing profitability and scale of its sugarcane biofuels business; in 2013 Shell canceled plans with Iogen Corporation for a commercial-scale plant in Canada—other past investments included Codexis, Virent, Energy Systems and HR Biopetroleum. Currently Shell is a player in the sugarcane ethanol business through Raizen (Brazil) and recently started up a cellulosic ethanol plant using sugarcane bagasse as feedstock; Chevron shelved plans back in 2010 after examining 100 different feedstocks, while ExxonMobil spent USD 100 million over 4 years on algae only to cancel the program; Valero pulled out USD 232 million investment in Mascoma wood to ethanol plant in Michigan, besides past investments in VeraSun, Renew Energy, Terrabon, Qteros, Zeachem, Solix; DuPont is one of the companies that, according to EPA, could start producing substantial quantities of cellulosic biofuels, it is a strategic matter to its biotechnology business—DuPont Industrial Biosciences. They have invested in a cellulosic ethanol plant located in IA, USA (Jessen, 2013; NEXTSTEPS, 2014; Provine, 2014; Lane, 2015a).

PwC has carried out a market survey on second-generation biofuels, 250 respondents were reached from different organizational levels as follows: (1) 31% CEO, Owner, President, or Chairman; (2) 19% Senior Vice President, Vice President, Executive Vice President, or Board Director; (3) 31% Director or Manager; (4) 19% Other. The question related to funding source was answered by 164 respondents ($n = 92$ USA and $n = 72$ rest of the world); the results are presented in Fig. 1.3. The majority of respondents have very high confidence in the long-term success of biobased chemicals. Chemicals and renewable fuels remain top end products. USA respondents were less confident about cellulosic ethanol than the rest of the world. The top four regions considered most important for companys' overall growth prospects in 1-year horizon were, the

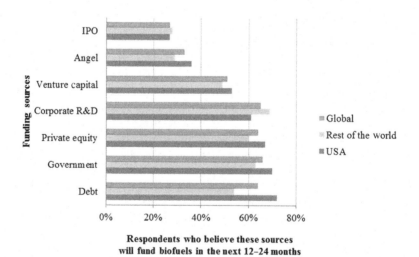

FIGURE 1.3   Short-term funding sources for second-generation biofuels. *Adapted from PWC, (2014) 2nd Biofuels Market Insights Survey. 2 pp. Available from:* http://www.pwc.com/us/en/technology/publications/biofuels-market-survey.jhtml.

United States, Europe, Brazil and Asia (companys' headquarters, 54% in the United States, 18% in Europe). The confidence of respondents was also measured focusing on what type of fuels could reach 1 billion gallons or more in global production volume by 2020, the results per fuel were: (1) 61% cellulosic ethanol; (2) 59% renewable diesel; (3) 52% biobased chemicals; (4) 48% aviation biofuels; (5) 35% military fuels; (6) 20% biobutanol; (7) 18% renewable gasoline; (8) 13% algae-based fuels. Long-term technology confidence outputs (most confident = 5) to the United States (rest of the world) were, cellulosic ethanol: 2.9 (3.4), drop-in fuels: 3.4 (3.2), biobased chemicals: 4.0 (4.0), and algae-based biofuels: 2.0 (2.0). Government and policy support were captured as the most significant industry concern, with US respondents being more concerned than the rest of the world (PWC, 2014). In general, the survey mapped a cautiously optimistic scenario.

Cellulosic ethanol market entrants have been considering transitional projects, also labeled as bolt-on or collocation, to gain experience with feedstock while using the existing infrastructure and supply chain to the largest extent possible. For example, Raizen's project in Brazil applying sugarcane bagasse as feedstock (NEXTSTEPS, 2014). Companies like DuPont, Beta Renewables, Abengoa, CTC, which at a glance can be considered as equity partners and/or technology providers most likely discuss internally and with potential business partners about cellulosic ethanol plant capacities keeping the eventual leveraging of the partner's structure available as part of the technological and economic analyses (eg, biomass availability and logistics, steam and electricity supply, human resources, technology maturity, performance guarantees, etc.). Globally, the first-generation business group's know-how is worth implementing on those projects, for instance, in Brazil sugarcane straw is already collected by sugarcane plants to fuel boilers and generate electricity, experience which can be transferred to cellulosic ethanol enterprises. The decision of choosing to apply that biomass (straw or bagasse) or replacing that by lignin as a fuel to boilers (steam to electricity) is also a matter to be part of business and technical discussions. Finally, plant capacity, technology offer, operational strategy (standalone, bolt-on or conservative incremental approach—small capacities), and economics have to be strategically evaluated with certain willingness to risk.

Cellulosic ethanol may have been oversold, at least early in the 2000s, regarding marketing readiness. After years of industry efforts, commercial-scale second-generation biofuel companies, using different technology platforms, are taking their first steps across the finish line. Commercial-scale plants with high potential of being successful follow a classic pattern—long-term research and methodical scale-up. The DuPont and Abengoa programs can be used to illustrate such a classic pattern. They have evolved gradually trough scales (bench, pilot, demonstration and commercial) having done many runs in demonstration plants prior to commercial projects. DuPont's journey to market with cellulosic ethanol is presented in Fig. 1.4. With a biofuels presence on three continents, Abengoa is an international biotechnology company—one of the largest ethanol

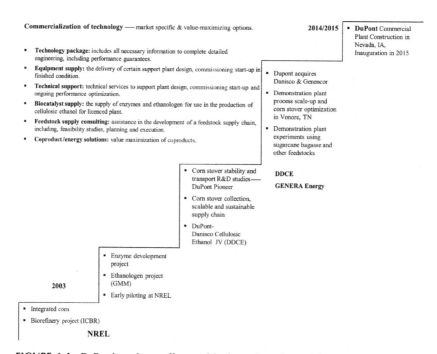

FIGURE 1.4 DuPont's scale up efforts and business dynamics. *Adapted from Provine, W.D., 2014. DuPont's Journey to Build a Global Cellulosic Biofuel Business Enterprise. In: Biomass 2014. Washington, DC. Available from:* http://www.energy.gov/eere/bioenergy/downloads/dupont-s-journey-build-global-cellulosic-biofuel-business-enterprise.

producers in the United States and Brazil, and the largest producer in Europe with a total of 867 million gallons of annual installed production capacity distributed among 15 commercial-scale plants in five countries (Breining, 2014; Provine, 2014; Standlee, 2014).

For the global biofuel industry, 2014 proved to be a leading-edge year. Five new commercial-scale cellulosic ethanol plants started production, three in the United States, one in Brazil and one in Italy. As the technology matures its expected capital investment in plants with capacity of 100 million gallon/year (~380 million liters/year), a practical limit imposed by the logistics of moving and storing up to 2000 metric tonnes/day of low density cellulosic biomass. Many plants are being planned in Brazil, Italy, Macedonia, Slovakia, China and the United States (Brown et al., 2015). Table 1.2 summarizes the commercial cellulosic ethanol plants based on the sugar platform (AEC, 2012; Brown et al., 2015; Licht, 2015). For projects based on the thermochemical platform, five companies are planning to open or are currently in operation: INEOS Bio (wood and vegetable waste), Enerkem (municipal waste), Lanzatech (planned 2015—wood waste and miscanthus), CoolPlanet (planned 2016—yellow pine)

TABLE 1.2 Commercial-Scale Cellulosic Ethanol Projects Using Sugar Platform

Company	Location	Opening Year	Capacity/ Year Million Gallons (in Liters)	Feedstock	Cost (Millions)
Abengoa Bioenergy	USA	2014	25 (95)	Corn stover, wheat straw and grasses	USD 504
Beta Renewables	Italy	2014	20 (76)	Wheat straw	€90
GranBio	Brazil	2014	22 (83)	Sugarcane, straw	USD 195
POET-DSM	USA	2014	20 (76)	Corn stover and cobs	USD 250
Raizen (logen technology)	Brazil	2015	10 (38)	Sugarcane bagasse	USD 100
DuPont	USA	2015	25 (95)	Corn stover	USD 225
Inbicon	USA	2015	10 (38)	Wheat straw	–
Mascoma	Canada	delayed	20 (76)	Hardwood and pulpwood	USD 200
Inbicon	Denmark	2016	20 (76)	Wheat straw	–
Beta Renewables	USA	2016	20 (76)	Energy grass	USD 200
Canergy	USA	2017	25 (95)	Energy cane	–
Energochemica	Slovakia	2017	16.5 (62)	Wheat and rapeseed straw	–
Abengoa	Brazil	2017	16.9 (64)	Sugarcane bagasse	–
DuPont-Ethanol Europe	Macedonia	–	–	Agricultural residues	–
MG Chemicals and Anhui	China	–	20 (76)	Agricultural residues	–
Progetti Italia	Italy	–	–	Agricultural residues	–

Source: Data from AEC, 2012. Cellulosic Biofuels: Industry Progress Report 2012-2013. 35 pp. Available from: http://ethanolrfa.3cdn.net/d9d44cd750f32071c6_h2m6vaik3.pdf; Brown, T.R., Brown, R.C., Estes, V., 2015. Commercial-scale production of lignocellulosic biofuels. Chem. Eng. Prog., March, 62–64; Licht, F., 2015. Cellulosic biofuels—the next five years. Int. Sugar J. March, 182–184.

and Fulcrum Sierra Biofuels (planned 2016—municipal waste), those plants presented a capital investment per annual gallon of approximately USD 13.50 (USD 3.60/liter), total capacity considering all five thermochemical plants is 49 million gallon/year (185 million liters/year), Enerkem is in Canada and the rest of the plants are in the United States. The same approach is planned for the carboxylate platform, only one plant is planned, but delayed, ZeaChem plant is also located in the United States and it was planned to 25 million gallon/year capacity with capital investment per annual gallon of about USD 16.00 (~USD 4.00 per annual liter), with agricultural residues and hybrid poplar to be used as feedstock. Using this platform, nearly all biomasses can be converted into carboxylate salts, which can be chemically transformed into a wide variety of chemicals and hydrocarbon fuels (Brown et al., 2015; Holtzapple et al., 2015). Unsuccessful examples of advanced biofuels, unfortunately, have also been reported. Range Fuels sold its 4 million gallon plant (15 million liters) to Lanzatech from New Zealand for around USD 5 million some years ago, after it had invested no less than USD 356 million in building the facility. KiOR, which owns a USD 190 million plant producing drop-in fuels using a thermochemical approach, came into serious problems like difficulties achieving target product yields. The company generated several thousand RINs in the course of 2013 (a few hundred thousand gallons), but this was not sufficient to keep the plant running (Licht, 2015).

Back to 2030 projections, fully processing the available agriculture residues, the United States could double and Brazil more than triple annual production volumes through cellulosic ethanol, assuming 2015 first-generation ethanol forecast as a baseline. Provocatively and cautiously illustrating, it would need approximately 750 plants spread all over each country to integrally access the biomass available, assuming a base case plant capacity of 95 million liters (eg, DuPont and Abengoa projects). Putting things into perspective, a significant share of Brazilian plants crush 3–5 million tonnes of sugarcane/season, with annual ethanol production ranging from 255 to 425 million liters/year (season). Currently, there are about 400 first-generation ethanol plants in Brazil and 200 in the United States which were built over recent decades. In the period of euphoria, 2008/2009, over 30 plants opened in Brazil (~USD 200 million/plant), about 80 plants have however closed since 2008. The integral processing of agriculture residues (17.5% availability) would require the construction of 50 cellulosic ethanol plants per year in the coming 15 years. Moreover, it is reasonable to estimate that it takes about 12–15 months to build each of those facilities, not to mention the capital and long-term policies needed to ensure an aggressive implementation. The opportunity is definitely huge as well as an implementation challenge. Although current projects have already partially released that "bio-well," it is reasonable to expect the industry will evolve in a more modest pace than would be required to fully access the 2030 projected biomass.

Projects will evolve analyzing case-specific economic feasibilities including biomass availability and logistics as an important factor to design plant capacity. For example, currently, a sugarcane ethanol plant which crushes 3.5 million tonnes produces about 315 million liters of ethanol. For every tonne of cane harvested there are 140 kg of dry basis straw available, part of that material has to stay in the field for agronomic reasons. Assuming 50% to be used to cellulosic ethanol, the amount to be processed would be 245,000 metric tonnes, so taking into account the yield of 400 L/dry tonne—2030 optimistic premise, the annual production would be 98 million liters, potentially a 95-million-liter project. In another view, if the same plant was currently under analysis (technology-neutral), it would or at least should consider about or even less than 300 L/dry tonne for the first years of operation, so the annual production would reach over 70 million liters. Taking into account the learning curve and technology advances, it is reasonable to design the plant assuming production will gradually increase towards 95–98 million liters. There are, of course, other important variables in that equation which have to be deeply analyzed to support business decisions, for instance, biomass availability and supply cluster, use of straw to fuel boilers and release of bagasse to cellulosic ethanol, cellulosic ethanol stillage usage as fertilizer, electricity generation using lignin stream, etc. The GranBio (83 million liters) cellulosic ethanol project in Brazil is a business case which most likely evaluated all those factors in their discussions prior to building their plant collocated at Caeté site (Carlos Lyra group), a traditional sugarcane plant located in the northeast of Brazil which has an annual sugarcane crushing capacity of 2 million tonnes (total of 280,000 tonnes of biomass). The rough sugarcane biomass calculations mentioned play the role of illustrating the rational feedstock versus plant capacity. It is important to note that Brazilian first-generation ethanol plants do not follow a standard on size, besides some are ethanol-dedicated and others sugar and ethanol. In addition, although all generate their own electricity, not all are energetically optimized to better use biomass potential, typically pursuing to generate a surplus to be sold to the grid. Such an optimization rationale is also valid to cellulosic ethanol analysis to find the better strategy to implement a plant with an economically and sustainable product portfolio, electricity, ethanol and sugar. National industry associations like UNICA (Brazil), RFA (USA), and ePure (Europe) typically keep databases on installed capacity features.

Looking at projected capacities in the United States it is clear they will not, by far, be sufficient to meet the US mandate. It is estimated that cellulosic ethanol and other advanced biofuels at the end of 2020 will be around 700 million liters per year, below 200 million gallons. Cellulosic ethanol can achieve more than 600 million liters. None of the recently built cellulosic ethanol plants has operated long enough to prove its economic viability and the plants were not only set up to serve the RFS. They were also built to prove that the operator's production process is a promising option. Companies like DuPont, POET-DSM JV or INEOS also target to license their production technology. Apart from the current plants' economic performance, the RFS mandates that they must be high enough to bring some new capacity online (Licht, 2015). In spite of the intense US oil and gas industry lobbying against the mandates, they actually want ethanol, but not that much. Ethanol is a desirable as an additive due to its high octane content, so the oil industry's perspective on optimal mix of ethanol for this purpose is roughly 10%. This demand is and would be attended by corn ethanol installed capacity even if there was no RFS. On the other hand, the "gallon in, gallon out" policy effect can jeopardize the first-generation industry, under a business-as-usual scenario. Corn ethanol blending could continue up to its 15 billion gallon cap—2022 statutory RFS, and cellulosic fuel that was produced would be blended in addition to that volume. Without a real demand created by RFS, corn ethanol producers would rely on the 10% additive market. However, in that 10% limited market context, cellulosic ethanol gets preference on the line and consequently every gallon coming online would replace a gallon of corn ethanol and cause impact on the first-generation business. Once large volumes of cellulosic ethanol most likely will not be produced in the short term, the current corn ethanol industry installed capacity will continue to be dedicated to attend the additive market needs. That market and mandate dynamics may disengage the first-generation industry from the effort. This may impact other future partnership launches like current POET-DSM, for which the major attractiveness for DSM was POET's fleet of 27 corn ethanol plants, which could provide significant demand for the new cellulosic technology and the potential for long-term profitability (Fitzpatrick, 2015).

In Europe, some changes in policy may come, but their potential impact on advanced biofuels is yet unclear; the last compromise was for conventional biofuels of 7% for 2020 while the remaining 3% should be met with advanced biofuels, e-cars and others. That could mean European demand for advanced biofuels of 14 billion liters by 2020. Only Italy offered a legal perspective for cellulosic ethanol for the years to 2022. With technology ready for commercial deployment, and funding available, it seems all that is left to boost demand for advanced biofuels is EU-wide implementation of cellulosic ethanol blending targets and a clear, ambitious energy framework post-2020 (Licht, 2015; Downey, 2015; The Economist, 2015; ePure, 2015).

In Brazil, GranBio and Raizen started to pave the cellulosic ethanol road using sugarcane bagasse and straw as feedstocks. Research and development into advanced biofuels is ongoing in the state-owned energy company Petrobras. BNDES has been very supportive to industrial biotechnology projects as well as to overall sugarcane sector focusing productivity gains and diversification. The BNDES effort on cellulosic ethanol is a significant move towards the building of advanced biofuel policies. As reported in their study, considering volume needed to attend current blending increase (25–27%), about 1 billion liters is required, if a policy existed where 10% of the volume had to be cellulosic ethanol, the first plants would be incentivized enough to distribute their product in the local market instead of targeting USA or Europe mandates. Besides, a gradual increasing in cellulosic ethanol blending may boost new investments in the sugarcane sector. Other alternatives were also explored by BNDES: subsidize the consumption, tax incentives and investment support to new plants, long-term funding to applied research, regulatory improvements, and human capital development (BNDES, 2015). Brazil's sugarcane ethanol industry has a challenging future. First of all, it needs to recoup from past negative impacts caused by government misconceptions, global financial crisis, and weather problems. Meanwhile, it needs to keep the peripheral view highly active to be able to capture value from new opportunities like cellulosic ethanol and other biochemicals. It is also part of the context, to stay aware of eventual global government stepbacks to concentrate on using cheaper oil and gas products which could be detrimental to the global environment and blur import barriers, particularly between South America, the United States and Europe. For example, for the eventual natural-gas-based ethanol production in the United States, long-term investment decisions are also needed and are dependent of national LNG and oil export posture (Traylen, 2014; Light, 2014; Mielnik, 2015). Electricity business dynamics may also influence cellulosic ethanol projects in Brazil. Local consumption is estimated to grow 40% in the next 10 years. There are thermoelectric plants applying sugarcane bagasse and/or straw and other biomass sources (eg, wood chips) to fuel boilers. Currently, for a few sugarcane business groups, bioelectricity has been a "cash cow" contributing to overall business sustainability (ethanol, sugar, dry yeast, bioelectricity). In addition, government recently set up auction prices which may encourage investment growth in biomass-based thermoelectric facilities (Macedo and Sousa, 2010; Ruiz, 2015). Therefore, it is expected that bioelectricity relevance may increase in coming years, this implies a thorough analysis on thermodynamic (lignin material and/or bagasse and straw) and economic aspects towards a sustainable economic model to justify cellulosic ethanol and/or bioelectricity enterprises. A study of scenarios for the ethanol industry in Brazil was done in 2010 considering a horizon extending to 2020. One of the designed scenarios covered bioelectricity and pointed out, in an absence of incentives for cellulosic ethanol, bioelectricity may be the choice for sugarcane biomass value creation. In general, the scenarios indicated that the development of cellulosic ethanol on an industrial scale depends primarily on political factors and, in the second stage, on technical achievements (Raele et al., 2014). BNDES efforts to better understand cellulosic ethanol business dynamics and propose potential policy pathways are on a certain level convergent to that study.

Cellulosic ethanol is a long-term investment and global market conditions are essential for projects to evolve. Nonfood biofuel potential justifies this crusade. Stakeholders need to focus on making it happen, but it is not a matter of business as usual. Uncertainties on policies are echoed all over this chapter and it is possible to realize its impacts on investment pace, clearly the global ethanol industry needs investment to continue to expand, but without a solid policy pillar, at least for a certain period of time, the risk willingness will be deteriorated. The sustainability of first-generation ethanol business plays a significant role on fostering companies to consider substantial investment moves towards advanced biofuels and biochemicals.

FINAL REMARKS

The biofuels industry has already proved its capability to positively contribute to environmental concerns and the economy through billions of dollars being annually invested, green job creation and reduction of GHG emissions. Policymakers have the challenge to progress on biofuel discussions towards policies which can reduce market uncertainties and foster industry to increase the investment pace. Governments have a significant role on that journey as well as society on improving the perception value on biofuel and biochemical overall benefits.

Cautiously, it can be stated there is a certain "leakage of innovation" ongoing in the United States and Europe which favors the global ethanol industry. Companies and governments have funded hundreds of millions of dollars on research programs to develop technologies and produce biofuels from nonfood crops, so return on investment is expected. The two cellulosic ethanol plants operating in Brazil have been using foreign technologies. Although Brazil has a traditional first-generation ethanol industry, there is still not a national technology provider for cellulosic ethanol and/or other biomass "drop-in" fuels.

It is valid to notice that apparently in the United States and Europe there is a mismatch between innovation and manufacturing and/or energy policies. They may not be effectively taking the advantage of technologies through new plant investments. However, through companies they are already able to extract value from any progress in global biofuel policies which could accelerate the growth of second-generation biofuels. Brazil is not in the same technological level, which makes the country a technology importer. It is fair-minded to state that Brazil is gradually progressing, supported by BNDES sponsorship on strategic technological surveys and policy proposals as well as through local ethanol and biotechnology industry associations lobbying. The Brazilian government has also been providing funds to incentivize universities, research centers and companies to work on this gap.

The cellulosic ethanol participation in global production pursuing the foreseen three-digit billions of liters in demand is certainly dependent on policies, but also on technology reliability. The learning curve maturation has a significant role to boost business cases through a reduction of capital investment and manufacturing costs. Pioneers of this nascent industry currently need to subsidize the curve through their other businesses and/or loans. In Brazil, large risk willingness was observed in companies investing in cellulosic ethanol without previous experience in the biofuel industry (eg, GranBio).

Ethanol investors are familiar with the first-generation ethanol cost reductions as a consequence of learning-by-doing, apparently in second-generation ethanol a more aggressive dynamics boosted by close scientific support has been intrinsically assumed to give faster access to such reductions. Although some company failures were mentioned throughout the chapter, the recent global wave of second-generation biofuel plants can have a higher chance of success. They may quickly cross the "valley of death" taking into account the indirect learning and the claims of technology providers on being ready-to-market. Precompetitive collaboration out of proprietary knowledge borders could also contribute to accelerating the industry growth. Brazil is an important market to technology licensers as well as China. It is important to notice technology providers are not risk-shielded and usually they have to sign contracts with performance guarantee and intellectual property protection terms.

The first-generation installed capacity spread all around the world continues to be important to attend the near-future demand and some investments may be observed, like few corn ethanol plants in Brazil. As captured throughout this chapter some of those first-generation sites can be leveraged to collocate second-generation plants. In the United States, cellulosic ethanol mandate can foster the demand growth, but the rest of the world has still to advance in that matter. There are positive industry expectations regarding COP21 results. In general, the chapter allows realizing that there is no sign of a truce by the ethanol industry in the search for ways to achieve its growth potential, cellulosic ethanol plays a significant role in that process.

ACKNOWLEDGMENTS

The author wishes to thank his family for their understanding and patience throughout the writing of the chapter. He also would like to thank Professor Sergio Luiz Monteiro Salles-Filho for the invitation and encouragement to be part of this project.

REFERENCES

AEC, 2012. Cellulosic Biofuels: Industry Progress Report 2012-2013. 35 pp. Available from: <http://ethanolrfa.3cdn.net/d9d44cd750f32071c6_h2m6vaik3.pdf>.

AEC, April 15, 2015. Advanced Ethanol Council Letter to United States Senate Committee on Finance: Advanced Ethanol Council Comments on Tax Reform. Available from: <http://advancedethanol.net/wp-content/uploads/2015/04/AEC_tax_Senate2015WGcomments.pdf>.

Bake, W.J.D., Junginger, M., Faaij, A., Poot, T., Walter, A., 2009. Explaining the experience curve: cost reductions of Brazilian ethanol from sugarcane. Biomass Bioenergy 33, 644–658.

BETO, July 2014. Appendix D: 2012 Cellulosic Ethanol Success. Available from: <http://www.energy.gov/sites/prod/files/2014/07/f17/appendixD_july_2014_0.pdf>.

BNDES, 2011. PAISS-resultado da etapa de seleção dos planos de negócio. Available from: <http://www.bndes.gov.br/SiteBNDES/export/sites/default/bndes_pt/Galerias/Arquivos/produtos/download/paiss_planos_selecionados.pdf>.

BNDES, 2013. PAISS: BNDES/Finep Joint Plan to Support Industrial Technology Innovation for the Sugarcane Industry. Available from: <http://www.globalbioenergy.org/fileadmin/user_upload/gbep/docs/2013_events/GBEP_Bioenergy_Week_Brasilia_18-23_March_2013/4.8_MILANEZ.pdf>.

BNDES, 2014. A produção de etanol pela integração do milho-safrinha às usinas e cana-de-açúcar: avaliação ambiental, econômica e sugestões de política. Revista do BNDES, 147–208.

BNDES, 2015. De promessa a realidade: como o etanol celulósico pode revolucionar a indústria da cana-de-açúcar—uma avaliação do potencial competitivo e sugestões de política pública. Revista do BNDES, 237–294.

BNEF, 2012. Moving Towards a Next-Generation Ethanol Economy. Available from: <https://www.dsm.com/content/dam/dsm/cworld/en_US/documents/bloomberg-next-generation-ethanol-economy.pdf>.

Brandiwad, A., 2014. Bio-butanol: back to the future. Bioenergy Connection 3.1, 22–23.

Breining, G., 2014. The cellulosic biofuels odyssey. Bioenergy Connection 3.1, 11–19.

Brown, R.C., Capareba, S.C., 2015. Producing biofuels via the thermochemical platform. Chem. Eng. Prog. March, 41–44.

Brown, T.R., Brown, R.C., Estes, V., 2015. Commercial-scale production of lignocellulosic biofuels. Chem. Eng. Prog. March, 62–64.

Carmann, T.M., 2011. Biobutanol: profile of an advanced biofuel and its path to market. Int. Sugar J. 113, 94–100.

Downey, K., 2015. Breaking the vicious cycle. Biofuels International March/April, 28–29.

Dwyer, M., 2015. U.S. ethanol in a global market: implications for trade. In: Licht, F.O. (Ed.), Sugar and Ethanol Brazil. São Paulo Slides download available only for participants.

E4tech, 2015. Final Report for the European Commission Directorate—General Energy: From the Sugar Platform to Biofuels and Biochemical. Available from: <https://ec.europa.eu/energy/sites/ener/files/documents/EC%20Sugar%20Platform%20final%20report.pdf>.

Elabora Consultoria, 2014. World Directory of Advanced Renewable Fuels and Chemicals. Elabora Editora, São Paulo.

EPA, 2015. Environmental Protection Agency Proposes Fuel Standards for 2014, 2015 and 2016, and the Biomass-Based Diesel Volume for 2017. EPA-420-F-15-028. pp. 1–4. Available from: <http://www.epa.gov/otaq/fuels/renewablefuels/documents/420f15028.pdf>.

ePure, 2014. Renewable Ethanol: Driving Jobs, Growth and Innovation Throughout Europe: State of Industry Report. Available from: <http://www.epure.org/sites/default/files/publication/140612-222-State-of-the-Industry-Report-2014.pdf>.

ePure, 2015. Renewable Ethanol: Enabling Innovation and Sustainable Development: State of Industry. Available from: <http://www.epure.org/sites/default/files/publication/ePURE-State-of-the-Industry-2015.pdf>.

Erickson, B., Nelson, J.E., Winters, P., 2012. Perspective on opportunities in industrial biotechnology in renewable chemicals. Biotechnology Journal 7, 1–10.

EurObserver, 2014. Biofuels Barometer. Available from: <http://www.energies-renouvelables.org/observ-er/stat_baro/observ/baro222_en.pdf>.

Farina, E., 2014. Sustainability of Biofuel Production: The Brazilian Ethanol and the Private Sector's Vision. Available from: <http://www.iea.org/media/technologyplatform/workshops/brazilnov2014/SustainabilityofBiofuelProductionTheBrazilianEthanolandThePrivateSecotorsVision.pdf>.

Farina, E., 2015. Producer's Survival Strategies: the 2015/2016 Crop and Beyond? Available from: <www.unica.com.br>.

Farina, E., Viegas, C., Pereda, P., Garcia, C., 2010. Ethanol Market and Competition. pp. 230–258. Available from: <http://sugarcane.org/resource-library/books/Ethanol%20Market%20and%20Competition.pdf>.

Fitzpatrick, R., 2015. Cellulosic Ethanol Is Getting a Big Boost From Corn, for Now. 21 pp. Available from: <http://www.thirdway.org/report/cellulosic-ethanol-is-getting-a-big-boost-from-corn-for-now>.

Gevo, December 2014. Luverne Plant Update. Available from: <http://ir.gevo.com/phoenix.zhtml?c=238618&p=irol-newsArticle&id=2005089>.

Goldemberg, J., Nogueira, L.A.H., 2014. The evolution of the biofuel's R&D, science and technology. Bioenergy Connection 3.1, 37–41.

Goldemberg, J., Coelho, S.T., Nastari, P.M., Lucon, O., 2004. Ethanol learning curve—the Brazilian experience. Biomass Bioenergy 26, 301–304.

Golden, J.S., Handfield, R.B., 2014. Why Biobased? Opportunities in the Emerging Bioeconomy. Duke University, Poole College of Management, Washington, DC, 34 pp. Available from: <http://www.biopreferred.gov/files/WhyBiobased.pdf>.

Guo, M., Weiping, S., Buhain, J., 2015. Bioenergy and biofuels: history, status and perspective. Renewable Sustainable Energy Rev. 42, 712–725.

Hettinga, W.G., Junginger, H.M., Dekker, S.C., Hoogwijk, M., Mcaloon, A.J., Hicks, K.B., 2009. Understanding the reductions in US corn ethanol production costs: and experience curve approach. Energy Policy 37, 190–203.

Holtzapple, M., Lonkar, S., Granda, C., 2015. Producing biofuels via the carboxylate platform. Chem. Eng. Prog. March, 52–57.

Huenteler, J., Anadon, L.D., Lee, H., Santen, N., 2014. Commercializing Second-Generation Biofuels. Harvard Kennedy School, Cambridge, 22 pp. Available from: <http://belfercenter.ksg.harvard.edu/files/commercializing-2ndgen-biofuels-web-final.pdf>.

ICCT, 2013. Measuring and Addressing Investment Risk in the Second-Generation Biofuels Industry. Washington, DC. 46 pp. Available from: <www.theicct.org>.

IEA, 2013. Production Costs of Alternative Transportation Fuels: Influence of Crude Oil Price and Technology Maturity. Paris. 46 pp. Available from: <https://www.iea.org/publications/freepublications/publication/FeaturedInsights_AlternativeFuel_FINAL.pdf>.

IEA, 2014. Renewable Energy: Medium Term and Market Report 2014. OECD/IEA. pp. 1–13. Available from: <www.iea.org>. Downloaded in: April 2, 2015.

IRENA, 2015. Renewable Energy and Jobs Annual Review. Available from: <http://www.irena.org/DocumentDownloads/Publications/IRENA_RE_Jobs_Annual_Review_2015.pdf>.

ISO, 2014. Ethanol Year Book 2014. International Sugar Organization, London, 160 pp.

Jessen, H., August 2013. Valero pulls out of Mascoma's Michigan proposed plant project. Ethanol Producer Magazine Available from: <http://ethanolproducer.com/articles/10150/valero-pulls-out-of-mascomas-michigan-proposed-plant-project>.

Krieger, K., 2014. Biofuels heat up. Nature 508, 448–449.

Lane, J., August 2014a. Butamax Opens (Phase 1) Biobutanol Project in Minnesota, With Highwater Ethanol. Available from: <http://www.biofuelsdigest.com/bdigest/2014/08/20/butamax-opens-phase-1-biobutanol-project-in-minnesota-with-highwater-ethanol/>.

Lane, J., December 2014b. BNDES OKs $116M in Financing for Abengoa Cellulosic Ethanol Plant in Sao Paulo State. Available from: <http://www.biofuelsdigest.com/bdigest/2014/12/19/bndes-oks-116m-in-financing-for-abengoa-cellulosic-ethanol-plant-in-sao-paulo-state/>.

Lane, J., January 2015a. BP's Exit From Cellulosic Ethanol: The Assets, the Auction, the Process, the Timing, the Skinny. Available from: <http://www.biofuelsdigest.com/bdigest/2015/01/18/bps-exit-from-cellulosic-ethanol-the-assets-the-auction-the-process-the-timing-the-skinny/>.

Lane, J., May 2015b. EPA Slashes Biofuels Targets for 2014, 2015, 2016 Under Renewable Fuel Standard. Available from: <http://www.biofuelsdigest.com/bdigest/2015/05/29/epa-slashes-biofuels-targets-for-2014-2015-2016-under-renewable-fuel-standard/>.

Langeveld, J.W.A., Dixon, J., Jaworski, J.F., 2010. Development perspectives of the biobased economy: a review. Crop Sci. 50, S142–S151.

Law360, June 2010. An Introduction too Cellulosic Biofuel Waiver Credits. Available from: <http://www.andrewskurth.com/media/article/1536_An%20Introduction%20To%20Cellulosic%20Biofuel%20Waiver%20Credits.pdf>.

LCTPi, 2015. Advanced Biofuels and the Climate Challenge. pp. 25–28. Available from: <http://lctpi.wbcsdservers.org/thesolution/>.

Licht, F., 2015. Cellulosic biofuels—the next five years. Int. Sugar J. March, 182–184.

Light, M., June 2014. Natural Gas Based Liquid Fuels: Potential Investment Opportunities in the United States. Available from: <http://www.goldmansachs.com/our-thinking/our-conferences/north-american-energy-summit/reports/fff-natural-gas-based-liquid-fuels.pdf>.

Lorenz, P., Zinke, H., 2005. White biotechnology: differences in US and EU approaches? Trends Biotechnol. 23, 570–574.

Macedo, C.I., Sousa, E.L., 2010. Ethanol and Bioelectricity. UNICA, São Paulo, Available from: <http://www.globalbioenergy.org/fileadmin/user_upload/gbep/docs/2013_events/GBEP_Bioenergy_Week_Brasilia_18-23_March_2013/4.8_MILANEZ.pdf>.

Markestrat, 2014. A dimensão do setor sucroenergético: mapeamento e quantificação safra 2013/2014. Fundace Business School and FEA-RP. 45 pp.

Martins, F., Gay, J.C., 2014. Biofuels: From Boost to Bust? Bain & Company, 8 pp.

MEI/CNI, 2013. Bioeconomy an Agenda for Brazil. Available from: <http://arquivos.portaldaindustria.com.br/app/conteudo_24/2013/10/18/411/20131018135824537392u.pdf>.

Mielnik, O., 2015. Petróleo: o xisto mudou o mercado. Agroanalysis 35 (2), 29–31.

Mousdale, D., 2014. Ten top indicators for liquid biofuel use in the next decade. Biofuels Bioprod. Bioref. 9 (3), 302–305.

NAP, 2015. Industrialization of Biology: A Roadmap to Accelerate the Advanced Manufacturing of Chemicals—Pre Publication Copy. The National Academies Press, Washington, DC, 126 pp. Available from: <www.nap.edu>.

Nastari, P.M., 2015a. Setor sucroalcooleiro: visão de médio e longo prazos. Agroanalysis 35 (2), 20–21.

Nastari, P.M., 2015b. Os efeitos da redução da mistura de etanol na gasolina. Agroanalysis 35 (5), 22–23.

Neves, M.F., 2015. Etanol é a aposta para alvancar setor sucroenergético. Available from: <http://www.markestrat.org/na-midia/etanol-%C3%A9-aposta-para-alavancar-setor-sucroenerg%C3%A9tico>.

NextSTEPS, 2014. Three Route Forward for Biofuels: Incremental, Transitional and Leapfrog. Davis. 41 pp. Available from: <www.steps.ucdavis.edu>.

OECD, 2009. The Bioeconomy to 2030: Designing a Policy Agenda—Main Findings and Policy Conclusions. Available from: <http://www.oecd.org/futures/long-termtechnologicalsocietalchallenges/42837897.pdf>.

OECD/Food and Agriculture Organization of the United Nations, 2014. OECD-FAO Agricultural Outlook 2014. OECD Publishing., Available from: <http://dx.doi.org/10.1787/agr_outlook-2014-en>.

Peplow, M., 2014. Cellulosic ethanol fights for life. Nature. March, 152–153.

Provine, W.D., 2014. DuPont's journey to build a global cellulosic biofuel business enterprise. In: Biomass. Washington, DC. Available from: <http://www.energy.gov/eere/bioenergy/downloads/dupont-s-journey-build-global-cellulosic-biofuel-business-enterprise>.

Pugatch Consilium, 2014. Building the Bioeconomy: Examining National Biotechnology Industry Development Strategies. 74 pp. Available from: <http://www.abiquim.org.br/download/comunicacao/apresentacao/building_the_bioeconomy_pugatch_consilium_april_2014.pdf>.

PWC, 2014. 2nd Biofuels Market Insights Survey. 2 pp. Available from: <http://www.pwc.com/us/en/technology/publications/biofuels-market-survey.jhtml>.

Raele, R., Boaventura, J.M.G., Fischman, A.A., Sarturi, G., 2014. Scenarios for the second generation ethanol in Brazil. Technol. Forecast. Soc. Change 87, 205–223.

REN21, 2015. Renewables 2015: Global Status Report. Available from: <http://www.ren21.net/status-of-renewables/global-status-report/>.

Ruiz, E.T.N.F., 2015. O renascimento da bioeletricidade. Agroanalysis 35 (6), 29–30.

Serigati, F., Possamai, R., 2015. A Petrobras ainda é a solução? Agroanalysis 35 (3), 18–21.

Service, R.F., 2014. Cellulosic ethanol at last? Science 345 (6201), 1111–1112.

Shubert, C., 2006. Can biofuels finally take center stage? Nat. Biotechnol. 24 (7).

Sorda, G., Banse, M., Kemfert, C., 2010. An overview of biofuel policies across the world. Energy Policy 38, 6977–6988.

Souza, G.M., Victoria, R., Joly, C., Verdade, L. (Eds.), 2015. Bioenergy & Sustainability: Bridging the gaps SCOPE, Paris. 779 pp.

Standlee, C., October 2014. Abengoa—Cellulosic Ethanol Update. EESI. Available from: <http://www.eesi.org/files/Chris_Standlee_100614.pdf>.

The Economist, 2015. Biofuels Thin Harvest. Available from: <http://www.economist.com/news/science-and-technology/21648630-investment-biofuels-dwindling-and-scepticism-growing-thin-harvest>.

Traylen, D., 2014. Heading south? Biofuels International March/April, 46–50.

Tyner, W., 2012. Biofuels and agriculture: a past perspective and uncertain future. Int. J. Sust. Dev. World Ecol. 19 (5), 1–6.

UNCTAD, October 2014. The Global Biofuels Market: Energy Security, Trade and Development. Policy in Brief. n. 30.

UNEP/BNEF. Global Trends in Renewable Energy Investment 2015: Executive Summary. Available from: www.fs-unep-centre.org. Downloaded in: May 20, 2015.

UNICA, December 2011. UNICA applauds ending of ethanol import tariff. Ethanol Producer Magazine Available from: <http://www.ethanolproducer.com/articles/8449/unica-applauds-ending-of-ethanol-import-tariff>.

UNICA, 2014. Estimativa da safra 2014/2015: coletiva de imprensa. Available from: <www.unica.com.br>.

Urbanchuk, J.M., 2015. Contribution of the Ethanol industry to the economy of the united states in 2014. Doylestown. 13 pp. Available from: <http://ethanolrfa.3cdn.net/813c483a4451ed5411_yxm6i6ov7.pdf>.

USDA, 2015. Biofuels-Brazil Raises Federal Taxes and Blend Mandate: GAIN Report Number BR 15001. São Paulo. 2 pp.

WEF, 2014. Biotechnology Ecosphere: Round Table of Biorefineries, Biotechnology and Bioenergy in Brazil. São Paulo. 15 pp. Available from: <http://www3.weforum.org/docs/IP/2014/CH/WEF_CH_Brazil_Biotechnology_Ecosphere.pdf>.

Wyman, C.E., 2007. What is (an is not) vital to advancing cellulosic ethanol. Trends Biotechnol. 25, 153–157.

Wyman, C.E., Dale, B.E., 2015. Producing biofuels via the sugar platform. Chem. Eng. Prog. March, 45–51.

Xiaoguang, C., Khanna, M., Yeh, S., 2012. Stimulating learning-by-doing in advanced biofuels: effectiviness of alternative policies. Environ. Res. Lett. 7 13 pp.

Yacobuci, B.D., July 2013. Analysis of Renewable Identification Numbers (RINs) in the Renewable Fuel Standard (RFS). Available from: <https://www.fas.org/sgp/crs/misc/R42824.pdf>.

Youngs, H., 2014. The bioeconomy is everywhere. Bioenergy Connection 3.1, 5–10.

Zhang, L., June 2014. Biofuels and Biochemical Then and Now: Innovation Trends From Feedstocks to End Products. 14 pp. Available from: <http://info.cleantech.com/Biofuels--Biochemicals-White-Paper-2014_Biofuels--Biochemicals-White-Paper-Submit.html>.

Chapter 2

Ethanol Use in the United States: Status, Threats and the Potential Future

W.M. Griffin[1], B.A. Saville[2] and H.L. MacLean[2,3]
[1]*Department of Engineering and Public Policy, Carnegie Mellon University, Pittsburgh, PA,*
United States [2]*Department of Chemical Engineering and Applied Chemistry, University of*
Toronto, Toronto, ON, Canada [3]*Department of Civil Engineering, University of Toronto, Toronto,*
ON, Canada

INTRODUCTION

Biofuels refers to liquid, solid and gaseous fuels produced from biomass, though this chapter focuses only on liquid biofuels. Biofuels include ethanol produced from corn and sugarcane, ethanol produced from cellulosic biomass, diesel fuel produced from soybeans and hydrocarbon fuels produced from algae. Cellulosics and algae-based fuels are still in their infancy.

First-generation biofuels refer primarily to ethanol produced from corn or sugarcane and biodiesel produced from soybean, rapeseed or sunflower. These biofuels are currently produced at the commercial scale and make a non-negligible contribution to meeting global demand for transportation energy. Ethanol from corn is the dominant biofuel in the United States and the industry produced over 14 billion gallons in 2014 (RFA, 2015).

While ethanol can be blended with gasoline and used in the light-duty fleet, there are some notable differences between the two fuel types. Ethanol has two-thirds the energy density of gasoline, so three gallons of ethanol are needed to offset the energy content of two gallons of gasoline (Bailey, 1996), notwithstanding any differences in combustion efficiency. For conventional vehicles, ethanol blends are limited to 15% or less by volume, recently adjusted upward from 10% (EPA, 2010a). Higher blends, up to E85, can be used in flex-fuel vehicles (FFVs), though the US vehicle fleet has only an estimated 17.4 million FFVs as of Sep. 2015 (AFDC, 2015a,b), accounting for roughly 6–7% of the total light-duty fleet.

Global Bioethanol.

In the United States, a number of events have shaped the recent transportation fuels market and concomitantly ethanol production and consumption. The continuing development of legislative mandates at both the state and national levels, responses to changes in the market, technology development, economic impacts of the introduction of these fuels, and changes in nascent regulations (eg, Corporate Average Fuel Economy (CAFE) standards) all shape the industry.

The following provides a general description of the processes for producing ethanol in the US context, the legislative history, and current approaches. Ethanol is not used in an economic "vacuum." This chapter addresses many of the obstacles to increased ethanol use in a system where vehicle technology is changing relatively rapidly compared to the past and new fossil resources are being discovered and rapidly developed.

CORN ETHANOL

Given modest feedstock cost, the production cost of corn ethanol has historically been low, thanks to both limited capital costs and considerable revenue generated from the major coproduct, distillers' dried grains. Assuming a feedstock cost of $2 per bushel, a technoeconomic analysis performed by the US Department of Energy's National Renewable Energy Laboratory (NREL) estimated corn ethanol production costs less than $1 per gallon (McAloon et al., 2000), and a United States Department of Agriculture (USDA) survey of corn ethanol producers largely confirms the estimate, finding average variable costs between $0.90 and $1 per gallon (Shapouri and Gallagher, 2005). Corn ethanol produced in the United States has historically been cost-competitive with sugarcane ethanol produced in Brazil (Gallagher et al., 2006). More recently, Perrin et al. (2008) surveyed a group of seven recently constructed ethanol plants and contrasted their findings with those of the USDA. They found net operating costs averaged over the plants of about $1.29 per gallon. The authors cited "high-cost plants" having costs at $1.35 per gallon and "low-cost plants" at $1.24.

Recent changes in the price of corn have drastically impacted the production cost of corn ethanol. Fig. 2.1 shows monthly average corn prices from the start of 2001 through the end of 2011. While average prices stayed between $2 and $3 per bushel from 2001 to the middle of 2006, prices have since remained above $3 per bushel, climbing above $5 per bushel during parts of 2008 and again in 2011.

Corn ethanol yields are generally 2.7–2.8 gallons per bushel (Shapouri and Gallagher, 2005), so an increase in corn cost of $1 per bushel increases ethanol production cost by $0.35 per gallon, or over $0.50 per gallon gasoline equivalent. Thus, the difference between the $2 per bushel assumed in the NREL study (which was representative of prices before 2006) and the $5 per bushel exceeded in recent years translates to an increase in the production cost of over $1.50 per gallon gasoline equivalent. Kwiatkowski et al. (2006) specifically modeled the

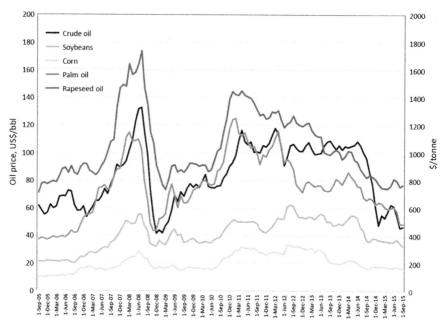

FIGURE 2.1 Ten-year monthly price history for corn, soybeans and crude, palm and rapeseed oils. *USDA (U.S. Department of Agriculture), 2015c. Feed Grains: Yearbook Tables.* <http:// www.ers.usda.gov/data-products/feed-grains-database/feed-grains-yearbook-tables.aspx> *(accessed 18.08.15.)* (USDA, 2015c).

sensitivity of ethanol price to changes in corn price. They found that if corn price increased by $1.80, then ethanol increased by $0.50. This is slightly lower than the back of the envelope calculation but still in the same range.

It is unclear exactly how much of the increase in corn prices is due directly to increases in corn ethanol production, but some studies have found that a portion of recent changes can be attributed to biofuels. In 2014, almost 38% of the harvested corn crop was used for ethanol production (USDA, 2015a) although some of this demand is offset by the return of distillers' grain to the animal feeds markets. The USDA predicts corn acreage to be between 88 and 90 million acres, with yields steadily increasing to 185 bushels/acre (USDA, 2015b) through 2024. Using this much corn would most likely support higher prices.

The issue of impacting commodity food prices arose with the rapid rise of corn prices from 2007 to 2009. Some studies found that biofuel production was responsible for 70% of the price rise (Lipsky, 2008) while others found no effect (Ajanovic, 2011). Since the rise, corn prices have declined, although not to the levels seen before the rise (Fig. 2.1). Interestingly, the price of corn, palm oil, rapeseed oil and soybeans has tracked the rise and fall in crude oil prices since 2005, even though the supply/demand balance for each of these commodities is likely different.

CELLULOSIC ETHANOL

Unlike first-generation biofuels like corn and sugarcane ethanol and biodiesel, second-generation biofuels like ethanol and liquid hydrocarbon (or "drop-in") fuels produced from cellulosic biomass are still largely under development, despite considerable research and investment. Two commercial-scale cellulosic ethanol plants are operating in the United States, owned by POET-DSM and Abengoa, although they are in the commissioning and start-up phase. Dupont also has a commercial-scale plant nearing completion. If successful these plants could provide about 80 million gallons of ethanol per year. Quad County has added the unit operation to convert corn fiber to cellulosic ethanol and produces 2 million gallons per year. Even with this start, the industry is at an early stage and there is considerable uncertainty related to many of the factors that will drive the prices and supplies of cellulosic fuels in the future, and subsequently, the impact that cellulosic fuels will have the success of the technology.

Despite these uncertainties, it is possible to identify a number of factors that will collectively determine the impact of cellulosic fuels. One of these factors is feedstock cost. Like corn and sugarcane ethanol, the cost of producing cellulosic ethanol and other cellulosic fuels is highly sensitive to feedstock costs. Sources of biomass considered as feedstocks for fuel production (or electricity production) include agricultural residues, hard- and softwoods, herbaceous energy crops and mill and urban wood residues. Agricultural residues include corn stover and wheat straw and represent the portions of the plant that remain once the crop itself is harvested. Although some residue must be left on agricultural land to prevent erosion and maintain soil carbon, the only costs to obtaining agricultural residues involve collection, transportation and storage, making agricultural residues a promising low-cost, near-term biomass feedstock. Perlack and Turhollow (2003) estimated delivered costs of $40–50 per dry ton of corn stover, though costs could be considerably higher or lower based on resource availability. Similarly, Graham et al. (2007) estimated that approximately 60 million Mg (66 million tonnes) of corn stover could be collected throughout the United States at costs of $30 per dry tonne or less. More recent estimates (Thomas, 2015; Tiller, 2015) suggest feedstock costs in the range of $80–115 per dry tonne.

Hard- and softwood biomass includes logging residues, forest thinnings, and short-rotation woody crops such as willow and hybrid poplar trees. Like agricultural residues, forest residues represent a currently available feedstock where the feedstock cost is composed primarily of collection, transportation and storage costs. Short-rotation woody crops grown in tree farms could increase the supply of woody biomass, but likely at a higher cost than forest residues. Residue collection and transport cost estimates vary greatly for different regions and terrains. For instance, Rummer et al. (2003) estimated costs of $40–95 per dry Mg ($36–86 per dry tonne) to cut and extract fuel reduction materials from gentle terrain, with costs increasing by about 20% for rolling terrain, though costs could potentially be significantly higher for mountainous regions in the western United States. Polagye et al. (2007) considered the production of wood pellets, bio-oil and methanol from thinnings throughout the state of Washington, estimating an average cost of $11 per Mg for cutting and skidding thinned biomass and an additional $6–8 to load, chip and debark biomass for use as a biofuel feedstock. Cost estimates in Polagye et al. are relatively low due to high stand density and favorable terrain throughout the thinned region. Eriksson and Gustavsson (2008) found that stumps and small roundwood produced from thinning operations could be removed at a cost of $70–125 per dry Mg. Thus, costs for wood residues may be higher or lower than costs estimated from crop residues given the significant variation in terrain and resource density.

Herbaceous energy crops refer primarily to perennial grasses such as switchgrass and *Miscanthus*. These crops may be grown on marginal or degraded cropland without requiring heavy pesticide or fertilizer application, but due to uncertainty regarding energy crop yields and cellulosic feedstock prices, it may be difficult for energy crops to make a significant contribution to meeting demand before a mature market for cellulosic biomass has developed. That being said, it may be possible to grow energy crops at modest production costs. Perrin et al. (2008) analyzed switchgrass production in commercial-scale fields and found average costs of $65 per Mg, which could fall below $50 per Mg when projected over a 10-year harvesting cycle. McLaughlin and Kszos (2005) estimated that a farm gate price of $44 and $52 per Mg would result in the production of switchgrass on 16.8 and 21.3 million hectares, respectively, throughout the United States. With projected yields of 9 Mg per hectare, switchgrass could account for over 150 million Mg of cellulosic feedstock per year at prices near $50 per Mg. Farmers will likely require a robust and reliable cellulosic feedstock market before investing heavily in energy crop production, but if that market exists, then energy crops have the potential to meet cellulosic feedstock demand at relatively low prices.

Urban wood waste and mill residues represent perhaps the most economically favorable near-term feedstock. Like agricultural and forest residues, these feedstocks represent waste streams with no production cost, with the added advantage of higher density and reduced collection costs compared to other residues. However, these waste streams may contain impurities whose removal increases the net feedstock cost, and some of these biomass sources are already used to provide energy for industrial operations.

Basic feedstock cost estimates are not the only part of the equation. Rather, the relationship between feedstock cost and supply is more relevant to the impact that cellulosic fuels will have on meeting future transportation energy demand than cost estimates alone. Perlack and Stokes (2011) examined the entire biomass resource potential for the United States under various farm gate prices and growth scenarios. They estimated that 0.9 billion dry tons of biomass could be available per year by 2022 and just over a billion tons by 2030 at $60 per ton. It is important to realize that this is biomass "at the roadside" and not with delivery and further processing costs required to be a readily utilizable feed-stock at the cellulosic biorefinery. Using a high-yield scenario, which impacts the agriculture waste and energy crop estimates, increased the 2022 estimate to 1.2–1.3 billion tonnes and the 2030 estimate to 1.4–1.6 billion tonnes. These values may overestimate actual future biomass availability, though, as some biomass sources may be too dispersed or too far from bioenergy processing facilities to make them economically viable, potentially stranding a significant amount of biomass due to high collection and shipping costs.

Haq (2002) examined the availability of biomass within the context of the US Energy Information Administration's (EIA's) National Energy Modeling System (NEMS), which is used to develop EIA's Annual Energy Outlook. Though the analysis was performed from the perspective of using biomass to generate electricity, feedstock supply projections are relevant for liquid fuels as well. By 2020, over 400 million dry tonnes of biomass are expected to be available annually at a price of $5 per million Btu, or around $80 per dry tonne. However, only urban wood waste and mill residues are expected to provide a significant amount of energy at prices below $2 per million Btu, with the total supply from these sources reaching 25 million dry tonnes at low prices.

Finally, it is likely that initial cellulosic feedstock costs will be higher than many of the estimates above. Those estimates are generally more representative of a mature bioenergy industry, and short-term feedstock costs may be higher. Hamelinck et al. (2005) modeled cellulosic ethanol in the short, medium and long term and estimated that cellulosic ethanol feedstock cost would likely drop from $75 per dry tonne to $50 per dry tonne as the industry matures. Cellulosic ethanol yields are expected to rise from around 70 gallons per dry tonne to close to 100 gallons per dry tonne of feedstock, so the difference of $25 per dry tonne in moving from a short-term to a long-term scenario corresponds to a difference of $0.25–0.35 per gallon of ethanol.

In the short term, cellulosic feedstock costs may not be a significant driver in lowering cellulosic ethanol prices. Farmers must be induced to provide these resources and thus, residues will likely be exploited first since they add revenue to that generated by the current crop choice and farming practices. The large-scale production of cellulosic energy crops means a shift for the farmer to new crops. This uncertainty is enhanced by the chicken and egg problem. There is a need for feedstock to grow the cellulosic fuels industry and there is a need for a robust cellulosic fuels industry to demand the feedstock. This will likely require government intervention to move past this early dilemma. Feedstock costs are likely to come down as the industry matures, and if feedstock costs decline significantly, it would likely decrease the price of cellulosic ethanol and liquid transportation fuels.

ALGAE AND PHOTOSYNTHETIC BACTERIA

The lay public and company marketing groups have been referring to organisms that can be used to capture light and convert the energy to biomass, oils or other products such as algae and/or microalgae. To limit confusion we will refer to potential eukaryotic algae and photosynthetic bacteria simply as algae. The strict scientific distinction is not necessary for this discussion.

Algae represent an alternative to food crops and terrestrial biomass for biological conversion of sunlight into useful energy. These organisms use sunlight to combine water and CO_2, trapping useful energy in the form of biomass, and potential energy products eg, oils, alcohols and hydrocarbons.

Algae may be ideal for production of bioenergy due in part to their simplicity. Algae are less complex than higher plants, but accomplish photosynthesis via similar mechanisms. The simpler structure supports faster growth rates and simpler nutrient requirements. As a result, algae have the potential to produce at least 30 times more energy per unit of land than terrestrial biofuel feedstocks (Sheehan et al., 1998). Meeting renewable fuel targets using algae would require drastically less land than meeting similar targets using terrestrial biomass. Furthermore, the land that is required for algae production can be non-agricultural, reducing the competition with food created by both first-generation biofuel feedstocks and energy crops.

Algae are capable of producing more valuable fuels than ethanol. They can produce oils and hydrocarbon fuels that are compatible with current production and transportation infrastructure, and this avoids the distribution and vehicle infrastructure hurdles facing other alternative fuels (eg, high-level ethanol blends).

One of the first concerted research efforts examining bioenergy production from algae was the Aquatic Species Program (ASP) operated by the US Department of Energy from 1978 to 1995. This program began with a focus on producing hydrogen from algae, but quickly transitioned to concentrate on production of liquid fuels. The ASP made a number of advances in molecular biology, genetic engineering and algae species collection, and demonstrated mass production of algae using an open pond system. The program was discontinued in 1995 due to budgetary constraints (Sheehan et al., 1998).

Recent petroleum prices, combined with concerns over the environmental impacts of first-generation biofuels, have led to a renewed interest in production of fuel from algae, though high production cost remains the major downside. Benemann and Oswald (1996) provides a detailed technoeconomic analysis of algal ponds used to convert CO_2 for bioenergy production. Previous analyses that were summarized in the study estimated costs between $140 and $180 per barrel of algal oil, given in 1994$. The updated analysis revised those already high estimates further upward, resulting in total costs of $160–265 per barrel of algal oil. The higher costs resulted from the cost of pure CO_2 (instead of using scrubbed flue gas from coal plants) and lower algal productivities. Adjusted for inflation, these cost estimates correspond to prices of $250–400 per barrel of oil, three to 10 times greater than recent prices for crude oil. The costs of producing oil from open algal ponds would need to drop considerably before attracting commercial investment.

The ASP concluded that alternatives to the open pond design tested in that program were less attractive for large-scale bioenergy production due to their higher cost requirements. However, recent advances in the mechanical design of closed photobioreactors and genetic engineering of photosynthetic organisms have made closed systems far more cost-effective. Robertson et al. (2011) outline a number of these advances in a project designed to produce biofuel from cyanobacteria. In contrast to most algae, which produce some combination of oils, carbohydrates and biomass during cell growth, cyanobacteria can be engineered to produce oil that is exported from the cell into the surrounding system. This avoids the need for costly biomass harvesting and oil extraction. The proposed closed-loop bioreactor prevents contamination from outside organisms, protecting the genetically engineered bacteria. While closed-loop bioreactors do incur increased capital costs compared to open pond systems, high productivity and continuous operation result in long-term cost projections as low as $20 per barrel of diesel and $0.60 per gallon of ethanol (Lane, 2015). This greatly improved productivity accounts for much of the vast difference between the open pond cost projections provided in Benemann and Oswald (1996) and the extremely low long-term projections for biofuel from cyanobacteria. Fig. 2.2 shows how the conversion efficiency of sunlight into liquid for cyanobacteria compares to both an open pond design and the theoretical maximum efficiency.

With the above in mind, production of biofuel from algae is further from commercialization than cellulosic biofuels from terrestrial crops. Both open pond (for production of bioenergy rather than nutritional products) and closed bioreactor designs need to overcome a number of technical obstacles before they can be demonstrated at commercial scale. Biofuel production from algae has greater potential than biofuel production from terrestrial crops due to the drastically different land requirements, but being so far from commercial scale production, cannot be reliably predicted to influence liquid fuel prices in the near to intermediate term.

BUTANOL

Recently, biobased butanol has attracted attention with BP and DuPont announcing plans to sell sugar-beet-derived butanol as a gasoline-blending component in the United Kingdom (Hess, 2006). This announcement in combination with reported research and development advances in butanol production and cited fuel property advantages of butanol compared to ethanol suggests butanol could become a potential alternative/additive to gasoline.

Butanol is not currently used as a fuel in North America. Recent testing, however, has found that butanol has several fuel property advantages compared to ethanol and may be a better blending component the ethanol. However, blending 16% of *n*-butanol in gasoline is the "legal equivalent" to an E10 blend (AFDC, 2015c).

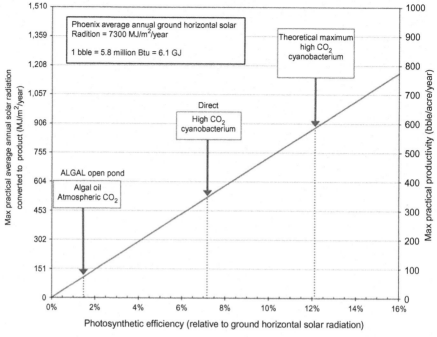

FIGURE 2.2 Photosynthetic efficiency and practical productivity for algae ponds with ambient CO_2, cyanobacteria with high CO_2, and the theoretical maximum conversion with high CO_2. *Adapted from Robertson, D., Jacobson, S., et al., 2011. A new dawn for industrial photosynthesis. Photosynth. Res. 107 (3), 269–277.*

Butanol has a higher energy density than ethanol, it is soluble in gasoline, is immiscible with water, so it can be distributed in existing pipelines either as a blend or neat.

There are two distinct approaches to producing butanol, developing and optimizing the traditional acetone, butanol, ethanol (ABE) fermentation or using genetic manipulations to place the needed metabolic machinery into a typical production organism. Fermentative production of butanol has a long history. During the first part of the last century, the ABE process was developed as a large-scale industrial fermentation of starch using *Clostridium acetobutylicum* to provide acetone for the munitions and airplane doping. By 1960, all production ceased due primarily to high feedstock (corn, molasses) costs and the large energy requirements for product recovery from dilute aqueous media. These increased costs made the ABE process uneconomic compared with petroleum production of the solvents (Jones and Woods, 1986).

Several recent advances have occurred in butanol production, including the development of the solvent-producing strain, *Clostridium beijerinckii* BA 101, with increased tolerance to solvents (Ezeji et al., 2003; Qureshi and Blaschek, 2001), and the development of advanced fermentation techniques and down-stream product recovery processes (Ezeji et al., 2007). Environmental Energy Inc. (EEI), subsequently ButylFuels LLC and now Green Biologics, patented a novel dual-stage process and claims the process significantly improves butanol yield and minimizes undesired byproducts (Ramey and Yang, 2004).

James Liao and colleagues (Atsumi et al., 2008a,b) at UCLA have developed an *Escherichia coli* strain incorporating genes from various clostridial species to produce a variety of alcohol related products, for example, butanol and isobutanol. The research laboratory has constructed a strain that produces isobutanol at 86% of theoretical yield. GEVO is currently producing isobutanol using this approach from corn starch at a converted ethanol mill in Luverne, MN. At this point the production is related to process development to support licensing efforts and chemical production for market development. Similarly, Green Biologics is converting an ethanol plant in Little Falls, MN, to produce *n*-butanol from corn grain.

Butamax has recently been granted a number of patents that have adopted a similar strategy to Liao except focusing on the ethanol-producing yeast *Saccharomyces*. It should also be noted that Jay Keasling at UC Berkeley and his colleagues have developed a *Saccharomyces cerevisiae* strain for the production of *n*-butanol. Professor Keasling is the intellectual driving force behind Amyris, so one might assume this technology will be commercially developed as well (C&E News, 2011).

Historically, starch and sugar feedstocks have been utilized for butanol production; however, like other biofuels, butanol could be produced from lignocellulosic biomass. During the last several decades, research has been conducted on this topic (Yu et al., 1984; Zverlov et al., 2006; Qureshi et al., 2007). Currently, however, there is no commercial-scale production of butanol from starch, sugar or lignocellulosic feedstocks and there are limited data on the potential processes. Thus, the commercial success of butanol is uncertain.

The theoretical yield of butanol is 0.41 g/g of glucose and ethanol is 0.51 g/g, suggesting that the cost per liter of ethanol will be less on a volumetric basis since feedstock cost is a large portion of the overall production costs for both processes. Butanol's solubility and energy density advantage more than off-set this yield disadvantage if these characteristics can lead to a price premium. However, in the current marketplace, even the energy difference between ethanol and gasoline is not recognized. Thus, at least in the short term butanol will need to compete on volumetric price, which is not likely considering the yield, density and separation energy disadvantages.

With this volumetric cost disadvantage compared to ethanol, initial production will likely occur in Brazil where there is low-cost sugar. Brazilian production will make butanol's other perceived advantages, blending characteristics and pipeline transmission, moot because the product will need to be shipped to market and blended, perhaps as ethanol is blended today.

CURRENT LEGISLATIVE INITIATIVES FOR BIOFUEL USE

Energy Policy Act of 2005 (RFS1)

The US Congress enacted the Energy Policy Act of 2005 to ensure secure, affordable and reliable energy. This act established the initial Renewable Fuel Standard (referred to as RFS1) and mandated 4 billion gallons of ethanol be used in 2006, rising to 7.5 billion gallons by 2012. This bill was superseded by the Energy Independence and Security Act of 2007 (EISA).

Energy Independence and Security Act 2007 (RFS2)

EISA is arguably responsible for the recent rapid rise in US biofuel production and use, particularly corn ethanol. The legislation established RFS2, which ultimately mandates the use of 36 billion gallons of biofuels by 2022 (EISA, 2007). RFS2 created targets for domestic consumption of biofuels, and developed a classification system distinguished mainly by lifecycle GHG emissions profiles; renewable biofuels (essentially corn ethanol), cellulosic biofuels (either as cellulosic ethanol or drop-in fuels) and other undifferentiated advanced biofuels. The required GHG reductions compared to gasoline (or diesel for biodiesel) are 20% for new corn ethanol capacity, 50% for advanced biofuels and 60% for cellulosic biofuels. Fig. 2.3 shows the biofuel production targets established by RFS2 for each of these fuel classifications for the years 2008–22. The act simply mandates increasing volumes of biofuel use annually out to 2022, subject to annual reviews by the Environmental Protection Agency (EPA). How ethanol is to be incorporated into the national fuel system is not specifically addressed, only that obligated parties are required to purchase the ethanol (or Renewable Identification Numbers (RINs)—discussed below). The implicit assumption is that ethanol would be blended in gasoline at appropriate levels in EPA-approved fuels (E10 and E85), and if need be, the EPA would approve higher blends as needed. This has been done in the case of E15, but there is some resistance to adopting this fuel. (See short discussion of blend walls at the end of this section.)

Mandated corn ethanol production reaches a maximum production of 15 billion gallons per year in 2015, and current corn ethanol production has the capacity to meet this demand (RFA, 2015). Targets for cellulosic and other advanced biofuels are more challenging. Little cellulosic biofuel is produced at the commercial scale, while the target for production increases to over 5 billion gallons in 2015 and subsequently to 21 billion gallons in 2022. Ethanol Producer Magazine estimates the total annual capacity of cellulosic ethanol production at 71 million gallons, and this includes pilot and demonstration facilities. Five facilities owned by Abengoa, Fiberight, INEOS, Poet–DSM and Quad County Cellulosic Ethanol account for approximately 65 million gallons of annual capacity (EPM, 2015).

The EPA was directed by EISA to set yearly volume requirements for RFS2 based on national interest that includes impacts "on the environment, eg, air

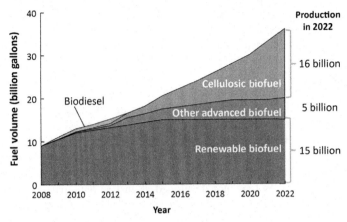

FIGURE 2.3   RFS2 volume requirements for renewable fuels through 2022 as mandated by Energy Independence and Security Act of 2007 (EISA, 2007).

TABLE 2.1 Renewable Fuel Volumes Proposed by EPA

	2014	2015	2016
	(Billion Gallons of Fuel)		
Cellulosic biofuels	0.033	0.106	0.206
Biomass-based diesel	1.63	1.7	1.8
Advanced biofuels	2.68	2.9	3.4

Source: Adapted from EPA (U.S. Environmental Protection Agency), 2015. EPA Proposes Increases in Renewable Fuel Levels. <http://yosemite.epa.gov/opa/admpress.nsf/0/4F4BB602AD5B51B98525 7E54004B6E5F> (accessed 18.08.15.).

quality, climate change, conversion of wetlands, ecosystems, wildlife habitat, water quality, and water supply," on energy security, rate of future production, impacts on infrastructure, costs to consumers, and impacts on "job creation, price and supply of agricultural commodities, rural economic development, and food prices" (EISA, 2007). Using these criteria, the EPA recently proposed volume standards for 2014 through 2016 (Table 2.1).

This volume-setting process is politically controversial, with EPA arguing the standards "deliver on Congressional intent to increase biofuels use, lower greenhouse gas emissions and improve energy security" and provide "a strong incentive for continued investment and growth in biofuels" (EPA, 2015). Opponents on both sides argue that the mandates are causing job losses in their respective industries, while others argue the impact on fuel cost and food costs to the American consumer. In some cases, the recent EPA volume requirements require the use of more fuel than is available in the market (US DC District Court, 2013). The proposed volumes for cellulosic ethanol also exceed the current capacity of the industry in both 2015 and 2016. These disagreements have resulted in numerous legislative attempts to eliminate the mandates for ethanol altogether (The Hill, 2015).

The current EPA-proposed volumes differ considerably from those originally proposed in EISA. Cellulosic biofuel volumes are substantially lower (1.75, 3 and 4.25 billion gallons for 2014, 2015 and 2016, respectively) by over a factor of 1000. The requirements for advanced biofuels have decreased slightly (3.75, 5.5 and 7.25, for 2014, 2015 and 2016, respectively), and for biomass-based diesel, the original standards did not specify volumes for these years. In general, cellulosic technologies have not developed as fast as anticipated by technology optimists. The nation has had unanticipated increases in CAFE standards. The increased efficiency in the fleet and the impact of the "Great Recession" on general economic activity resulted in the first decrease in light-duty vehicle fuel use in the United States in 2008 (a 3% decline). Fuel use has remained relatively flat since that time, and with increased incorporation of ethanol, the actual volume of the gasoline component has continued to decrease. Although refiners continue to make gasoline at close to prior levels due to refinery process investments, the reduced US market for gasoline resulted in the United States becoming a net exporter of gasoline in early 2010, for the first time since 1949.

The RIN is a number associated with a specific type of biofuel salable under the renewable fuels program. RINs play an important part in monitoring the obligated parties' compliance. RINs allow the EPA to track the renewable fuels from production through blending. Under the program, any refiner that makes gasoline or diesel fuel or any importer of gasoline or diesel is assigned a renewable fuel obligation (RVO) based on fuel sales. The RINs are attached to each physical gallon of ethanol produced and transferred to the blender, where it is "detached" and used as proof to the EPA that the obligated party is meeting their RVO. RINs are tradable. An obligated party that overcomplies by purchasing more renewable fuel than required can sell RINs to those who do not.

Under RFS2, fuel blenders are not obligated parties but detach RINs. This fact has actually presented opportunities for independent fuel marketers to use and profit by selling RINs or ethanol blends, especially E85, for under market value. This gives them an advantage over producers that are also retailers (an integrated oil company for instance). The independent fuel marketers can purchase gasoline without a RVO. As they blend ethanol, they accumulate RINs, which can be sold and used to decrease their fuel sales price to the customer, or simply to pocket the revenue.

The impact of RFS2 on the overall price of liquid fuels depends mainly on whether or not cellulosic fuels will realize the low production costs projected for long-term scenarios. Production costs for cellulosic fuels will certainly be higher than production costs for first-generation biofuels and of the incumbent fossil fuels in the near-term. It is possible that production mandates in RFS2 will lead to investments in cellulosic ethanol that lead to significant production cost reductions through learning-by-doing. However, if cellulosic fuels production and cost reductions remain elusive, the production mandates in RFS2 will result in a domestic fuel blend that contains more advanced biofuel than the market would otherwise include because the volume of corn ethanol is capped. The resulting impact would be an increase in the price of liquid transportation fuels, due to the required inclusion of relatively high-cost biofuels (on an energy basis) into the fuel mix.

Many of the cost comparisons examining the impact of the increased use of renewable fuels before and after RFS2 have been done on a volumetric basis. Although the argument can be made that fuel mileage is more impacted by driving habits than by the energy reduction resulting from low-level biofuel blends, the overall fleet energy requirement is related to simply moving the mass of the vehicle, cargo, and passengers from point A to point B. Any reduction in the fuel energy/gallon will result in higher annual gallons of fuel use by the fleet if the fuel efficiency of the fleet remains the same. As the mandates for RFS2 increase, the reduction in energy per gallon in higher-level blends might be noticeable by the average driver. Fuel use has been decreasing on a per mile basis over the years since EISA was implemented, partly due to increased penetration of hybrid electric vehicles and increased CAFE standards.

Finally, the RFS has run into issues other than the lack of cellulosic fuels. The "blend wall" has constrained the ethanol volume that can be assimilated in the US fuel mix each year. Ethanol can be blended with gasoline up to 10% ethanol (E10) by volume, or 85% ethanol (E85). Intermediate blends are not permitted, with the exception of E15 for vehicles manufactured after 2001 (EPA, 2015). Automobile manufacturers have been hesitant to extend warranties for most US motor vehicles to cover higher ethanol blend use and retailers see the addition of another fuel as costly and potentially confusing to the consumer.

California Low Carbon Fuel Standard

A Low Carbon Fuel Standard (LCFS) represents a policy that reduces the carbon intensity of transportation fuels (on a lifecycle basis), but uses market mechanisms to do so in a cost-effective manner. An LCFS establishes a target carbon intensity value (given, for instance, in g CO_2e per MJ of fuel), and fuel blenders are required to sell fuel with an average carbon intensity value at or below that level. At the state level, California has currently instituted an LCFS with the goal of reducing the carbon intensity of transportation fuels by 10% by 2020 (Farrell and Sperling, 2007).

The approach is similar to cap and trade for the transportation sector, with the exception that the cap is placed on average carbon emissions per MJ of fuel sold, rather than on total emissions from all fuels. Both policies establish overall targets and provide flexibility for individual firms to collectively meet that target as efficiently as possible. In the case of the CA-LCFS, the regulated entities (fuel blenders) select the portfolio of fuel types and quantities that minimize cost, subject to the carbon intensity constraint. The carbon intensity targets can then be decreased over time to increase the carbon reductions achieved by the policy.

A national LCFS has the potential to greatly increase the cost of transportation fuels due to the lack of cost-effective low-carbon alternatives to petroleum. First-generation biofuels like corn and sugarcane ethanol have recently experienced considerable increases in feedstock cost, and the carbon reduction potential of these fuels has been challenged by high CO_2 emissions due to market-mediated effects like indirect land use change (ILUC) (see Section "Greenhouse Gas Emissions"). Advanced biofuels have not yet reached commercial-scale production, and must overcome significant obstacles before doing so. Overall, carbon emissions from transportation fuels are more costly to mitigate than emissions from other sectors, so any policy that requires emissions reductions, specifically from transportation fuels, will likely result in higher liquid fuel prices. Of course, oil prices could rise to a point where the renewable fuels could have the opposite affect. Due in large part to the expected increases in fuel cost that could result from an LCFS, such a policy seems unlikely at the national scale in the near future.

CAFE STANDARD

The Corporate Average Fuel Economy (CAFE) standard regulates the average fuel economy of new vehicles sold in the United States. CAFE standards work by establishing a required fuel economy for the entire fleet of new vehicles (in miles per gallon), with separate standards established for passenger cars and light trucks. Vehicle manufacturers are required to have a sales-weighted average fuel economy that meets or exceeds these fuel economy targets, or pay a penalty that is a function of both their total sales and fuel economy shortfall. Fig. 2.4 illustrates historical values of the CAFE standard for both passenger cars and light trucks, along with projections of potential future fuel economy standards (Shiau et al., 2009).

CAFE was originally enacted in 1975, in response to the Arab oil embargo. The fuel economy standards increased rapidly from the mid-1970s through the early 1980s, then were almost entirely flat through the mid-2000s.

To the extent that CAFE standards influence vehicle design decisions, CAFE policies can influence fuel consumption and price in a few ways. First, CAFE policies directly increase the fuel economy of the vehicle fleet, which directly leads to a reduction in fuel consumption. If the total vehicle miles traveled remains constant, then increasing the fuel economy of the fleet leads to a proportional decrease in fuel consumption, though some of those reductions in demand may be erased by the rebound effect. This demand reduction would directly lead to a reduction in price.

Second, CAFE standards may impact consumer decisions about whether or not to purchase a new vehicle. If CAFE standards are high, and are accompanied

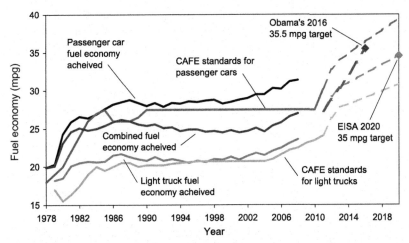

FIGURE 2.4   Historical CAFE standards, fleet fuel economy, and potential future policy trajectories. *Adapted from Shiau, C.-S.N., Michalek, J.J., et al., 2009. A structural analysis of vehicle design responses to corporate average fuel economy policy. Transp. Res. Part A Policy Pract. 43 (9–10), 814–828.*

by large penalties for violation, then they can influence vehicle design decisions in a way that steers the set of available vehicles away from the optimal set determined by the market based on consumer preferences, potentially leading to fewer new vehicle purchases. This secondary impact would primarily affect consumers looking for vehicles with low fuel economy and could therefore lead to additional increases in the fuel economy of the fleet. However, there may be some instances in which consumers decide not to purchase a new vehicle, and instead continue to drive old vehicles with poor fuel economy, leading to an increase in fuel consumption relative to the consumption from any new vehicle.

Shiau et al. (2009) modeled vehicle design responses to CAFE standards, assuming oligopolistic competition between firms and using a mixed logit model for the demand for different classes of vehicles. Shiau et al. find three regions in which CAFE standards have different impacts. For low values of the standard, market conditions lead to a higher average fuel economy than that specified by the policy, so firms will ignore the standard. For moderate values, the standard is treated as a binding constraint, since firms do not want to increase fuel economy, but would rather do so than pay the penalty. For high fuel economy standards, firms will ignore the standard and simply pay the penalty, where the critical point separating moderate and high standards (and therefore compliance and violation) is a function of the penalty for violation. Based on current models for producer and consumer preferences, Shiau et al. find that current CAFE standards are not binding, but that fuel economy is determined by market forces. Increasing the CAFE standard will have no effect unless the penalty for violating the standard is increased from the current penalty of

$55 per mpg. Furthermore, they find that fleet fuel economy is more sensitive to fuel prices than to the fuel economy standard and penalty established by the policy, so an explicit gasoline tax or carbon tax might have more influence on fuel consumption than fuel economy policies like CAFE.

EISA sets the goal to raise the fuel economy standard to 35 mpg by 2020, and probably most importantly, place light trucks under the same standard. Subsequently, the Obama administration entered into an agreement with 13 large automakers to increase the fuel economy standard to 54.5 mpg by 2025.

GREENHOUSE GAS EMISSIONS

Corn ethanol has traditionally been viewed as a fuel that achieves modest GHG emissions reductions relative to gasoline. A meta-analysis of corn ethanol lifecycle assessments (LCAs) (Farrell et al., 2006) found values for the GHG intensity of corn ethanol production ranging from 60 to 120 g CO_2 per MJ, with the majority of the variation resulting from variation in limestone application rate (to soil) and farm machinery energy use. Original analyses used in that study found modest average GHG reductions for corn ethanol production, with emissions from CO_2-intensive production exceeding those from gasoline (Farrell et al., 2006).

However, this highly cited and respected work was conducted before recent findings on emissions from land use change were publicized. Land use change emissions occur via two primary mechanisms, direct and indirect. Direct land use change (DLUC) emissions occur when land is converted from a native ecosystem into land used for agricultural production or between shifts in crops on currently productive land. The former conversion can result in a release of soil carbon and the release of carbon stored in existing vegetation. The total released carbon may be large in comparison to the carbon "saved" by displacing fossil fuels with biofuels (Fargione et al., 2008). The emissions associated with the latter conversion are considered to be small and in some instances might be negative, for instance if corn crops were displaced by switchgrass production. EISA does not permit bringing new land into production to meet its fuel mandates.

ILUC emissions also (like DLUC) refer to emissions from the conversion of land from a native ecosystem, but in the case of ILUC, land conversion occurs as a result of market-mediated impacts. For instance, if a bushel of corn is diverted from meeting food or other demands to being used as a feedstock for ethanol production, that diversion could slightly increase crop prices and induce land conversion to help meet preexisting grain demands. Some studies indicate that corn ethanol may be twice as carbon-intensive as gasoline once ILUC emissions are taken into account (Searchinger et al., 2008) estimated at 104 g CO_2e/MJ, but others strongly refute this. Numerous studies have shown values ranging from 19 to 65 g CO_2e/MJ (for a review see Khanna and Crago, 2012). The most likely scenario is one in which GHG reductions are possible under certain circumstances, whereas equivalent or higher carbon intensities may occur under different conditions, mainly impacted by the magnitude of consequential/indirect effects.

There are some drawbacks to the types of economic equilibrium models used to generate the ILUC estimates. Equilibrium models for predicting ILUC impacts rely on forecasts of complex global agricultural market forecasts that are impacted by considerable uncertainties on both supply and demand sides. Furthermore, the response of these markets to increased biofuel production needs to be precisely modeled on a spatial level, since land use changes can have drastically different carbon impacts based on the regional characteristics of the land converted to cropland. Future land use requirements also depend heavily on agricultural yields, which may continue to improve, especially for high feedstock prices. Issues involving the allocation of environmental impacts to biofuel coproducts (a problem faced by all LCAs of biofuels) are relevant to ILUC analyses as well.

The range of potential land use change estimates suggests considerable uncertainty related to the modeling of the phenomenon. Only recently has there been an attempt to understand the uncertainties related to using the general equilibrium models that provide the estimates beyond simple bounding. Plevin et al. (2015) used Monte Carlo analysis to propagate uncertainty analysis through the Global Trade and Analysis Project (GTAP) model. For corn ethanol they found that crop yield elasticities contributed about 50% of the variance observed in the mean emissions of 33 g CO2e/MJ, which ranged from 18 to 55 for a 95% confidence interval. The authors pointed out that there is considerable disagreement surrounding these values and thus were modeled using a wide distribution resulting in its contribution to the variance. Other fuel pathways were shown to have similar profiles.

Despite these drawbacks, land use change emissions present a serious challenge to the GHG benefits of corn ethanol. Recent analyses have settled on values for ILUC emissions considerably less than those originally presented in Searchinger et al. (2008), which estimated the contribution of ILUC emissions to be 100 g CO_2e per MJ of fuel. For instance, EPA settled on values of 40–55 g CO_2e per MJ for international land use change, depending on assumptions for the time frame and discount rate used in the analysis (EPA, 2010a,b), and the California Air Resources Board used a value of 30 g CO_2e per MJ in their analysis of corn ethanol emissions (CARB, 2009a). Though these estimates are considerably less than those originally presented in Searchinger et al. (2008), ILUC emissions still represent a large increase from non-ILUC emissions estimates for corn ethanol. A probabilistic analysis by Mullins et al. (2011) synthesized estimates for ILUC emissions, along with estimates for other uncertain LCA parameters, and generated estimates ranging from 50 to 250 g CO_2e per MJ for the carbon intensity of corn ethanol. Thus, based on a range of values published in the scientific literature, including ILUC emissions would likely make corn ethanol more carbon-intensive than gasoline, with the potential for a very large increase in emissions.

Greenhouse gas emissions from sugarcane ethanol production are generally estimated to be less than those from corn ethanol, due in part to the fact that sugarcane bagasse can be burned at the ethanol refinery to provide enough power to more than meet the energy requirements of the refinery. Without including emissions from land use change, an LCA by Macedo et al. (2004) estimated a carbon intensity value of only 11 g CO_2e per MJ of fuel, representing a reduction of almost 90% compared to gasoline. The California Air Resources Board (CARB, 2009b) generated estimates ranging from 12 to 27 g CO_2e per MJ for sugarcane ethanol produced in Brazil, depending on harvesting practices and the coproduct credit for electricity generation. These emissions estimates represent significant reductions relative to both gasoline and corn ethanol. Brazilian sugarcane is considered an advanced biofuel for the RFS2 mandate.

Those reductions could be challenged by emissions from land use change. Land use change emissions are particularly relevant in Brazil, where hypothetical indirect conversion of tropical rainforest due to the biofuel production could incur a "carbon debt" that takes over 300 years to repay through displaced fossil fuels (Fargione et al., 2008). However, estimates of land use change emissions resulting from production of sugarcane ethanol, and specifically ILUC emissions are highly uncertain. Estimates of sugarcane-induced land use change emissions are impacted by many of the same uncertainties discussed above in reference to corn ethanol, making the resulting estimates highly uncertain.

Cellulosic fuels have historically been viewed as a very low carbon alternative to fossil fuels. There are differences among the GHG impacts of cellulosic fuels produced from different feedstocks, but cellulosic fuels are generally estimated to have lifecycle GHG emissions well below those of gasoline or corn ethanol. Schmer et al. (2008) estimated that cellulosic ethanol produced from switchgrass offers a 94% reduction in GHG emissions relative to gasoline. Tilman et al. (2006) stated that lifecycle GHG emissions of biofuel production could even be negative if produced from high-diversity grassland biomass, where carbon sequestered in the soil during feedstock growth is greater than carbon released during biofuel production. Emissions from land use change could significantly change those estimates. For instance, Searchinger et al. (2008) estimated that ILUC emissions would cause cellulosic ethanol produced from switchgrass to have 50% higher GHG emissions than gasoline. An analysis performed for the California Air Resources Board of cellulosic ethanol produced from farmed trees estimated a much smaller but still significant value of 18 g CO_2e/MJ for land use change emissions (CARB, 2009c).

However, the impact that land use change emissions will have on fuel prices is entirely dependent on a mechanism for valuing emissions. Corn ethanol subsidies in many states within the United States are not tied to any specific level of emissions, so including very large estimates of ILUC emissions would not impact subsidies received by the fuel. The RFS2 does set a carbon intensity threshold for a fuel to qualify as a renewable biofuel, but those targets are not tied directly to any financial incentives.

ILUC emissions could potentially become important for a situation in which the United States enacted legislation valuing carbon emissions, either in the form of a carbon tax or a cap-and-trade mechanism. However, for near-term carbon prices (which almost certainly would be less than $50 per tonne CO_2), the cost of ILUC emissions may be small in comparison to the difference between ethanol and gasoline prices, even when assuming large values for ILUC emissions (Kocoloski et al., 2009).

ILUC emissions is a very important factor in the case of a carbon valuation policy established for only the transportation sector, such as the CA-LCFS (Farrell and Sperling, 2007). Neither comprehensive carbon legislation nor a transportation sector-specific carbon policy seem likely in the near term, so emissions from ILUC are expected to have little impact on the price of liquid fuels during this time frame; while in the intermediate term one could see enactment of carbon legislation, but initial values of a carbon tax or carbon emission permits are likely to be low. Only in the long-term, when high carbon prices or a policy specific to the transportation sector could come into play, would ILUC emissions have a significant impact on the price of liquid fuels.

Land use change emissions are expected to be less of an issue for cellulosic fuels than for corn and sugarcane ethanol. Unlike corn and sugarcane ethanol, use of cellulosic feedstocks for ethanol production does not *necessarily* compete with other demands. Cellulosic fuels can be produced from agricultural and forest waste streams, using biomass for which there is little to no current demand. Cellulosic energy crops can potentially be grown on marginal or degraded lands that may not be suitable for production of food/feed crops. Doing so may decrease yields or increase fertilizer and energy requirements relative to producing them on more fertile land, but could also mitigate grain price impacts that lead to ILUC. Only energy crops grown on fertile cropland are likely to directly compete with food/feed crops and induce significant amounts of land conversion.

Furthermore, the impact that ILUC findings have on the price of cellulosic fuels is a function of carbon valuation policy. For reasonable, near-term values of a carbon tax or carbon emissions permit, land use change emissions for cellulosic ethanol produced from energy crops will have a small impact. A policy that regulates GHGs only from the transportation sector would result in a large impact from land use change emissions, but only for fuel produced from energy crops that compete with food crops. Thus, the net effect of land use change emissions on cellulosic fuels may be a slight increase in price, but will more likely have a fairly small effect.

FUEL PRODUCTION IN THE UNITED STATES

Biochemical Ethanol Production Processes

The main process for producing ethanol is fermentation. Fermentation is a mature technology that is unlikely to change drastically in either cost or performance in the near future, although operators are always looking at new potential (the idea of a biorefinery or dry mills separating corn oil to produce biodiesel is an example). Other aspects of the biofuel production chain, including feedstock production, reducing energy requirements at refineries, and improvements in extracting fermentable sugars from cellulosic biomass, if improved, may drive the production costs of biofuels downward.

Liska et al. (2009) outlined a number of recent changes in corn production practices and refinery operations that improve corn ethanol production efficiency and decrease lifecycle GHG emissions. If public policies value GHG emissions (carbon tax, cap and trade), the decreases could translate into financial advantage for corn ethanol over liquid hydrocarbon fuels. Improvements in feedstock production practices and reduction of fossil energy inputs during corn ethanol production lead to lower production costs for corn ethanol.

Sugarcane ethanol production does not occur in the United States but is an important source of ethanol to meet the RFS2 and CA-LCFS regulations. Sugarcane ethanol production costs have the potential to drop in the near future due to improvements in feedstock production and refinery operations. Increased mechanization in sugarcane harvesting may decrease feedstock costs and increase the total biomass removed from the field. Mechanical harvesting is being legally phased-in throughout the state of São Paulo over the next 20 years, increasing the biomass available to be delivered to ethanol refineries (and used to generate electricity) by as much as 30% (Goldemberg et al., 2008). Increasing the electricity that is sold back to the grid improves the profitability of sugarcane ethanol refineries and drives down ethanol production costs.

Sugarcane bagasse frequently contains more energy than required to meet the demands of the refinery. Many sugarcane ethanol refineries can take advantage of this energy through the use of high-efficiency boilers and selling excess electricity returned to the grid. Refineries not connected to the grid will have more than enough process energy and frequently feature inexpensive, low-efficiency boilers to provide the needed process energy and dispose of what would then be considered a waste stream. When combining increased bagasse availability from mechanized harvesting with improvements in boiler efficiency, the result is the potential for increased electricity generation, greater profitability and lower sugarcane ethanol production costs.

Alternatively, excess bagasse could be used as a feedstock for cellulosic ethanol production. Over 70% of electricity generation in Brazil comes from hydroelectricity, so local generation is already cheap and clean, and additional generation from bagasse may have limited economic and environmental benefits. Producing cellulosic ethanol from bagasse could generate more value than producing electricity at the plant level, since plants could take advantage of existing fermentation and distillation capacity, as well as feedstock and fuel transportation infrastructure. Assuming future reductions in cellulosic ethanol production cost, coproduction of cellulosic ethanol from bagasse alongside sugarcane ethanol production could reduce the costs of both fuels and go a long way to helping the United States to meet the demands of RFS2.

Cellulosic ethanol has greater room for improvement than either of the first-generation ethanol fuels simply because facilities are just entering commercial-scale operations and have not experienced the "learning by doing" that can improve efficiency and reduce costs. Compared to corn and sugarcane ethanol, cellulosic fuels are expected to be capital-intensive. A comparative analysis by BBI Biofuels Canada (2010) estimated capital costs of $1.8 and $2.5 per gallon capacity for corn and sugarcane ethanol, respectively, and over $7 per gallon capacity for cellulosic ethanol. The large capital investment requirement represents one of the primary reasons that cellulosic ethanol has not yet contributed to meeting transportation energy demand.

Capital costs for cellulosic fuels do, however, have the potential to drop significantly in the future. An NREL report (Wooley et al., 1999) modeled a number of futuristic cases for cellulosic ethanol, projecting that capital cost requirements could drop to around $4.5 per gallon ethanol capacity in the short term, and could fall to as low as $1.8 per gallon ethanol capacity based on long-term improvements in both cost and fuel output. Similarly, Hamelinck et al. (2005) estimated capital costs falling from around $8 per gallon ethanol capacity in the near term to around $3 per gallon ethanol capacity in the long term.

Significant reductions in the required capital investment such as the ones projected by Wooley et al. (1999) and Hamelinck et al. (2005) would translate into large reductions in cellulosic ethanol production cost. Improvements in the performance of conversion technologies will also drive down cellulosic ethanol production costs. Much of this improvement would likely come in the form of higher efficiencies in converting the cellulose and hemicellulose in biomass into fermentable sugars. In an NREL analysis of futuristic scenarios, Wooley et al. (1999) estimated that ethanol yields could rise from 68 gallons per dry tonne of biomass up to 112 gallons of ethanol per dry tonne of biomass. It should be mentioned that the upper bound is virtually at the theoretical limit and would be highly dependent on feedstock composition. Analysis presented in Hamelinck et al. (2005) is not quite as optimistic, but still projects ethanol yield of 95 gallons per tonne of biomass for a long-term scenario.

Unfortunately, it is unclear whether or not the cellulosic ethanol industry will ever gain the experience necessary to realize some of these cost reductions. Despite considerable research and both public and private investment, conversion technologies for cellulosic ethanol have yet to be proven at the commercial scale for any length of time, although there are a number of pilot facilities and "production facilities" capable of producing millions of gallons of ethanol that have been built. Start-up has been difficult at this point.

Alternative fuel production pathways that convert biomass into liquid hydrocarbon fuels (ie, "drop-in" fuels) using the Fischer–Tropsch (FT) process are in development, and may reach commercial viability. Depending on costs, these pathways could compete with cellulosic ethanol for feedstock, providing downward pressure on liquid fuel prices, perhaps preventing cellulosic ethanol from making a significant contribution to meeting transportation energy demand.

At the same time, due to their longer history of commercial operation (on non-biomass feedstock), these FT processes are less likely to deliver meaningful cost reductions over time.

Should cellulosic ethanol approach commercial viability, capital cost reductions that generally occur as industry gains experience are likely to reduce production costs of cellulosic fuels and drive the price of liquid fuels downward, especially in the intermediate and long term.

Thermochemical Biofuel Production Process

In addition to the biological conversion of cellulosic biomass to ethanol, biofuels may also be produced through thermochemical conversion. Thermochemical conversion processes can yield a wide variety of fuel types, including hydrogen, diesel, gasoline, ethanol, methanol and mixed alcohols. Two of the most widely examined and mature thermochemical conversion methods are gasification of biomass to produce syngas and pyrolysis to produce bio-oil. In both cases, additional steps are required to upgrade or convert the end product into a fuel that can be used by the transportation sector.

Gasification converts carbonaceous materials into synthesis gas, or syngas, containing carbon monoxide, hydrogen, carbon dioxide and methane. Biomass gasification is more challenging than gasifying either coal or natural gas due in part to the high alkali content of biomass, which can lead to fouling problems (Huber et al., 2006). Once biomass has been gasified, it can be used to produce a wide variety of end products. Spath and Dayton (2003) examine different conversion pathways and end products for biomass-derived syngas, summarized in Fig. 2.5. Syngas can be converted into either diesel or gasoline via the FT process discussed above, or converted into hydrogen via the water–gas shift reaction. Alternatively, it can be converted into methanol, which can either be used directly as a fuel or converted into ethanol (Enerkem, 2015).

Production of either ethanol or drop-in fuels from cellulosic biomass through thermochemical processes possesses many of the same characteristics as biochemical production of cellulosic ethanol. Very competitive production costs are possible in the long term, but a number of technical and economic hurdles will likely prevent significant increases in production in the near term. Phillips et al. (2007) examined production of ethanol from forest biomass using indirect gasification and estimated a production cost of slightly more than $1 per gallon for a mature industry. Foust et al. (2009) compared biological and thermochemical conversion of biomass to ethanol. While thermochemical conversion produces less ethanol per ton of feedstock than biological conversion, thermochemical conversion can produce higher alcohols as coproducts, which reduces the thermochemical production cost by an estimated $0.10 per gallon compared to the biological conversion pathway.

Like cost estimates for biological production of ethanol from cellulosic biomass, the above cost estimates represent those achievable by a mature industry

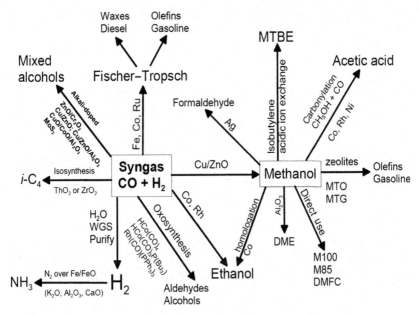

FIGURE 2.5   Potential syngas conversion processes and products (Spath and Dayton, 2003).

and may never be realized. But biofuel production via gasification has the potential for higher overall fuel yields (due to the fact that biochemical conversion cannot unlock the energy stored in lignin for fuel production) and can produce fuels that are more compatible with current transportation infrastructure. As a result, it may have a higher ceiling than biochemical conversion, and could become the primary conversion process for cellulosic biomass if progress in biochemical conversion remains slow.

Alternatively, drop-in fuels can be produced via pyrolysis. Pyrolysis involves heating biomass while restricting the availability of oxygen, and produces a combination of hydrocarbons known as bio-oil or bio-crude.

Pyrolysis has advantages over biochemical conversion. Pyrolysis requires much less time; fast pyrolysis can convert biomass into bio-oil in a matter of seconds, orders of magnitude faster than the time required to unlock and ferment sugars required in biochemical ethanol production. Conventional pyrolysis, which shifts end products toward solid fuels, may take hours, but is still faster than biochemical conversion. It also avoids the need to purchase and dispose of acid/catalyst required for pretreatment in certain cellulosic ethanol production processes (Ritter, 2011). On the other hand, bio-oil produced by pyrolysis is not identical to fossil crude oil, and most of the oils produced by pyrolysis (and an alternative, but most expensive, method, liquefaction) are low-grade, with considerable water, acids and tars that require upgrading before they can be blended with gasoline or diesel. Like hydrocarbon fuels produced via the

FT process, these "drop-in" fuels cannot literally be dropped in to petroleum supplies without being upgraded.

In the near-term, biofuel produced through pyrolysis is expected to cost considerably more than fossil alternatives. Huber et al. (2006) estimated costs for refined bio-oil produced via fast pyrolysis that range from $140 to $170 per barrel when adjusted for inflation. Pyrolysis may still be an attractive fuel conversion technology compared to gasification or hydrolysis and fermentation, due in part to limited capital costs, which only represent 15–20% of the total production cost (Huber et al., 2006). However, production costs must still decrease considerably in order for pyrolysis of biomass to produce fuels that are cost-competitive with petroleum-derived fuels.

FINAL REMARKS

The United States has a robust corn ethanol industry that is mainly driven through two regulatory programs; the California Low Carbon Fuels Standard and the national Renewable Fuels Standard. Ethanol is used in the United States as low-level blends (E10, E15) and as E85. Currently, cellulosic ethanol production is in its infancy and its success is still in question. As production increases, for the short term, cellulosic feedstock costs and the need for learning by doing for this new process will drive cellulosic ethanol prices. The regulatory approaches will necessarily change as states adopt programs to address climate change, while there is little appetite for this at the federal level and there continues to be an effort to rescind the national Renewable Fuels Standard.

A number of events have shaped the recent transportation fuels market and thus, ethanol production and consumption. The responses to changes in the market by consumers and the incumbent fuel producers, technology development in vehicles, economic impacts of the introduction of these fuels, and changes in nascent regulations (eg, CAFE standards) all shape the industry. Although not discussed here, the development of oil and gas production from shale, which is currently reducing petroleum costs and natural gas price to the point where new initiatives are being pursued to adopt liquefied natural gas and compressed natural gas fuels for transportation, will all impact future renewable fuel use.

REFERENCES

AFDC (Alternative Fuels Data Center), 2015a. Alternative Fuel Station Locator. <http://www.afdc.energy.gov/locator/stations/results?utf8=✓&location=&fuel=E85&private=false&planned=false&owner=all&payment=all&radius=false&radius_miles=5&e85_has_blender_pump=false> (accessed 07.09.15.).

AFDC (Alternative Fuels Data Center), 2015b. Flexible Fuel Vehicles. <http://www.afdc.energy.gov/vehicles/flexible_fuel.html> (accessed 07.09.15.).

AFDC (Alternative Fuels Data Center), 2015c. Biobutanol. <http://www.afdc.energy.gov/fuels/emerging_biobutanol.html> (accessed 18.08.15.).

Ajanovic, A., 2011. Biofuels versus food production: does biofuels production increase food prices? Energy 36, 2070–2076.

Atsumi, S., Hanai, T., Liao, J.C., 2008a. Non-fermentative pathways for synthesis of branched-chain higher alcohols as biofuels. Nature 451, 86–89.

Atsumi, S., Cann, A., Connor, M., Shen, C., Smith, K., Brynildsen, M., et al., 2008b. Metabolic engineering of *Escherichia coli* for 1-butanol production. Metab. Eng. 10, 305–311.

Bailey, B.K., 1996. Performance of ethanol as a transportation fuel. In: Wyman, C.E. (Ed.), Handbook on Bioethanol: Production and Utilization Taylor and Francis, Washington, DC, pp. 36–60.

BBI Biofuels Canada, 2010. Biofuel Costs, Technologies and Economics in APEC Economies, Final Report, December 2010. Available from: <http://www.biofuels.apec.org/pdfs/ewg_2010_biofuel-production-cost.pdf>.

Benemann, J.R., Oswald W.J., 1996. Systems and Economic Analysis of Microalgae Ponds for Conversion of CO_2 to Biomass. Final Report. PBD: March 21, 1996, 214 pp.

CARB (California Environmental Protection Agency Air Resources Board), 2009a. Detailed California-Modified GREET Pathway for Corn Ethanol. Preliminary Draft as Part of the Low Carbon Fuel Standard Regulatory Process, February 27, 2009. Available from: <http://www.arb.ca.gov/fuels/lcfs/022709lcfs_cornetoh.pdf>.

CARB (California Environmental Protection Agency Air Resources Board), 2009b. Detailed California-Modified GREET Pathways for Brazilian Sugarcane Ethanol: Average Brazilian Ethanol, With Mechanized Harvesting and Electricity Co-Product Credit, With Electricity Co-Product Credit. Preliminary Draft as Part of the Low Carbon Fuel Standard Regulatory Process, September 23, 2009. Available from: <http://www.arb.ca.gov/fuels/lcfs/092309lcfs_cane_etoh.pdf>.

CARB (California Environmental Protection Agency Air Resources Board), 2009c. Detailed California-Modified GREET Pathway for Cellulosic Ethanol From Farmed Trees by Fermentation. Preliminary Draft as Part of the Low Carbon Fuel Standard Regulatory Process, February 27, 2009. Available from: <http://www.arb.ca.gov/fuels/lcfs/022709lcfs_trees.pdf>.

C&E News, 2011. Amyris hits the market. Chem. Eng. News 89 (49), 26–28.

EISA (Energy Independence and Security Act of 2007), 2007. Library of Congress. Available from: <http://www.gpo.gov/fdsys/pkg/PLAW-110publ140/pdf/PLAW-110publ140.pdf>.

Enerkem, 2015. Exclusive Thermochemical Process. <http://enerkem.com/about-us/technology/> (accessed 22.08.15.).

EPA, 2010a. Renewable Fuel Standard Program (RFS2) Regulatory Impact Analysis. Assessment and Standards Division, Office of Transportation and Air Quality, U.S. Environmental Protection Agency, Agency USEP, Washington, DC.

EPA (U.S. Environmental Protection Agency), 2010b. EPA Announces E15 Partial Waiver Decision and Fuel Pump Labeling Proposal, October 13, 2010. Available from: <http://www.epa.gov/otaq/regs/fuels/additive/e15/420f10054.pdf>.

EPA (U.S. Environmental Protection Agency), 2015. EPA Proposes Increases in Renewable Fuel Levels. <http://yosemite.epa.gov/opa/admpress.nsf/0/4F4BB602AD5B51B985257E54004B6E5F> (accessed 18.08.15.).

EPM (Ethanol Producer Magazine), 2015. <http://www.ethanolproducer.com/plants/listplants/US/Existing/Sugar-Starch/> (accessed 18.08.15.).

Eriksson, L.N.S., Gustavsson, L., 2008. Biofuels from stumps and small roundwood-Costs and CO_2 benefits. Biomass Bioenergy 32 (10), 897–902.

Ezeji, T.C., Qureshi, N.M., Blaschek, H.P., 2003. Production of acetone, butanol and ethanol by *Clostridium beijerinckii* BA101 and in situ recovery by gas stripping. World J. Microbiol. Biotechnol. 19, 595–603.

Ezeji, T.C., Qureshi, N., Blaschek, H.P., 2007. Bioproduction of butanol from biomass: from genes to bioreactors. Curr. Opin. Biotechnol. 18, 220–227.

Fargione, J., Hill, J., et al., 2008. Land clearing and the biofuel carbon debt. Science 319 (5867), 1235–1238.

Farrell, A.E., Plevin, R.J., et al., 2006. Ethanol can contribute to energy and environmental goals. Science 311 (5760), 506–508.

Farrell, A., Sperling D., 2007. A Low-Carbon Fuel Standard for California, Part 2: Policy Analysis, eScholarship Repository.

Foust, T., Aden, A., et al., 2009. An economic and environmental comparison of a biochemical and a thermochemical lignocellulosic ethanol conversion processes. Cellulose 16 (4), 547–565.

Gallagher, P., Schamel, G., Shapouri, H., Brubaker, H., 2006. The international competitiveness of the U.S. corn-ethanol industry: A comparison with sugar-ethanol processing in Brazil. Agribusiness 22 (1), 109–134.

Goldemberg, J., Coelho, S.T., et al., 2008. The sustainability of ethanol production from sugarcane. Energy Policy 36 (6), 2086–2097.

Graham, R.L., Nelson, R., et al., 2007. Current and potential U.S. corn stover supplies. Agron. J. 99 (1), 1–11.

Hamelinck, C.N., van Hooijdonk, G., et al., 2005. Ethanol from lignocellulosic biomass: techno-economic performance in short-, middle- and long-term. Biomass Bioenergy 28 (4), 384–410.

Haq, Z., 2002. Biomass for Electricity Generation. U.S. Energy Information Administration Electricity Analysis Reports. <http://www.eia.gov/oiaf/analysispaper/biomass/>.

Hess, G., 2006. BP and DuPont to make biobutanol. Chem. Eng. News 84 (26), 9–10.

Huber, G.W., Iborra, S., et al., 2006. Synthesis of transportation fuels from biomass: chemistry, catalysts, and engineering. Chem. Rev. 106 (9), 4044–4098.

Jones, D.T., Woods, D.R., 1986. Acetone-butanol fermentation revisited. Microbiol. Rev. 50 (4), 484–524.

Khanna, M., Crago, C.L., 2012. Measuring indirect land use change with biofuels: implications for policy. Annu. Rev. Resour. Econ. 4 (4), 160–183.

Kocoloski, M., Griffin, W.M., et al., 2009. Indirect land use change and biofuel policy. Environ. Res. Lett. 4 (3), 034008.

Kwiatkowski, J., McAloon, A., Taylor, F., Johnston, D., 2006. Modeling the process and costs of fuel ethanol production by the corn dry-grind process. Ind. Crops Prod. 23 (3), 288–296.

Lane, J., 2015. Joule Says "Will Go Commercial in 2017": Solar Fuels on the Way. Biofules Digest. <http://www.biofuelsdigest.com/bdigest/2015/03/23/joule-says-will-go-commercial-in-2017-solar-fuels-on-the-way/>.

Lipsky, J., 2008. Commodity Prices and Global Inflation. Council on Foreign Relations, New York, NY.

Liska, A.J., Yang, H.S., et al., 2009. Improvements in life cycle energy efficiency and greenhouse gas emissions of corn-ethanol. J. Ind. Ecol. 13 (1), 58–74.

Macedo, I.C., Leal M., et al., 2004. Assessment of Greenhouse Gas Emissions in the Production and Use of Fuel Ethanol. Prepared for the Government of the State of São Paulo, April 2004.

McAloon, A., Taylor F., et al., 2000. Determining the Cost of Producing Ethanol From Corn Starch and Lignocellulosic Feedstocks. NREL/TP-580-28893.

McLaughlin, S.B., Kszos, L.A., 2005. Development of switchgrass (Panicum virgatum) as a bioenergy feedstock in the United States. Biomass Bioenergy 28 (6), 515–535.

Mullins, K.A., Griffin, W.M., et al., 2011. Policy implications of uncertainty in modeled life-cycle greenhouse gas emissions of biofuels. Environ. Sci. Technol. 45 (1), 132–138.

Perlack R.D., Stokes B.J., 2011. U.S. Billion-Ton Update: Biomass Supply for a Bioenergy and Bioproducts Industry. ORNL/TM-2011/224. Oak Ridge National Laboratory, Oak Ridge, TN, 227 pp.

Perlack, R.D., Turhollow, A.F., 2003. Feedstock cost analysis of corn stover residues for further processing. Energy 28 (14), 1395–1403.

Perrin, R., Vogel, K., Schmer, M., Mitchell, R., 2008. Farm-scale production cost of switchgrass for biomass. BioEnergy Res. 1, 91–97.

Phillips, S., Aden, A., et al., 2007. Thermochemical Ethanol via Indirect Gasification and Mixed Alcohol Synthesis of Lignocellulosic Biomass. Report Number NREL/TP-510-41168, 132 pp.

Plevin, R.J., Beckman, J., Golub, A.A., Witcover, J., O'Hare, M., 2015. Carbon accounting and economic model uncertainty of emissions from biofuels-induced land use change. Environ. Sci. Technol. 49, 2656–2664.

Polagye, B.L., Hodgson, K.T., et al., 2007. An economic analysis of bio-energy options using thinnings from overstocked forests. Biomass Bioenergy 31 (2–3), 105–125.

Qureshi, N., Blaschek, H.P., 2001. Evaluation of recent advances in butanol fermentation, upstream, and downstream processions. Bioprocess Biosyst. Eng. 24, 219–226.

Qureshi, N., Saha, B.C., Cotta, M., 2007. Butanol production from wheat straw hydrolysate using *Clostridium beijerinckii*. Bioprocess Biosyst. Eng. 30, 419–427.

Ramey, D., Yang S., 2004. Production of Butyric Acid and Butanol From Biomass. DE-F-G02-00ER86106. Technical Report Submitted to the US DOE Morgantown, Washington, DC.

RFA (Renewable Fuels Association), 2015. Statistics. <http://www.ethanolrfa.org/pages/statistics> (accessed 07.09.15.).

Ritter, S.K., 2011. Race to the pump: biofuel technologies vie to provide a sustainable supply of transportation fuels. Chem. Eng. News 89 (7), 11–17.

Robertson, D., Jacobson, S., et al., 2011. A new dawn for industrial photosynthesis. Photosynth. Res. 107 (3), 269–277.

Rummer, B., Prestemon J., et al., 2003. A Strategic Assessment of Forest Biomass and Fuel Reduction Treatments in Western States. U.S. Department of Agriculture, Forest Service, Research and Development, Washington, DC 2003(iii), 18 pp.

Schmer, M.R., Vogel, K.P., et al., 2008. Net energy of cellulosic ethanol from switchgrass. Natl. Acad. Sci. 105, 464–469.

Searchinger, T., Heimlich, R., et al., 2008. Use of U.S. Croplands for biofuels increases greenhouse gases through emissions from land-use change. Science 319 (5867), 1238–1240.

Shapouri, H., Gallagher, P.W., 2005. USDA's 2002 Ethanol Cost-of-Production Survey. Agricultural Economic Report Number 841.

Sheehan, J., Dunahay, T., et al., 1998. Look Back at the U.S. Department of Energy's Aquatic Species Program: Biodiesel From Algae; Close-Out Report. Report Number NREL/TP-580-24190, 325 pp.

Shiau, C.-S.N., Michalek, J.J., et al., 2009. A structural analysis of vehicle design responses to corporate average fuel economy policy. Transp. Res. Part A Policy Pract. 43 (9–10), 814–828.

Spath, P.L. and Dayton, D.C., 2003. Preliminary Screening—Technical and Economic Assessment of Synthesis Gas to Fuels and Chemicals With Emphasis on the Potential for Biomass-Derived Syngas. Report Number NREL/TP-510-34929, 160 pp.

The Hill, 2015. GOP Bill Would Repeal Federal Ethanol Mandate. <http://thehill.com/policy/energy-environment/245151-bill-would-repeal-federal-ethanol-mandate> (accessed 18.08.15.).

Thomas, S.R., 2015. Cellulosic biomass feedstock supply & logistics. In: Advanced Bioeconomy Feedstocks Conference, New Orleans, LA, June 9, 2015.

Tiller, K., 2015. Feedstock supply chains: from input cost to value driver. In: Advanced Bioeconomy Feedstocks Conference, New Orleans, LA, June 9, 2015.

Tilman, D., Hill, J., et al., 2006. Carbon-negative biofuels from low-input high-diversity grassland biomass. Science 314 (5805), 1598–1600.

USDA (U.S. Department of Agriculture National Agricultural Statistics Service (NASS)), 2015a. World Agricultural Supply and Demand Estimates. Available online from: <http://www.usda. gov/oce/commodity/wasde/latest.pdf> (accessed 18.08.15.).

USDA (U.S. Department of Agriculture), 2015b. USDA Agricultural Projections to 2024. U.S. Department of Agriculture, Washington, DC. Available online from: <http://www.usda.gov/oce/ commodity/projections/USDA_Agricultural_Projections_to_2024.pdf> (accessed 18.08.15.).

USDA (U.S. Department of Agriculture), 2015c. Feed Grains: Yearbook Tables. <http://www.ers.usda. gov/data-products/feed-grains-database/feed-grains-yearbook-tables.aspx> (accessed 18.08.15.).

US DC District Court, 2013. API vs. EPA. <https://www.cadc.uscourts.gov/internet/opinions.nsf/A57 AB46B228054BD85257AFE00556B45/$file/12-1139-1417101.pdf> (accessed 18.08.15.).

Wooley, R., Ruth M., et al., 1999. Lignocellulosic Biomass to Ethanol Process Design and Economics Utilizing Co-Current Dilute Acid Prehydrolysis and Enzymatic Hydrolysis Current and Futuristic Scenarios. Report Number NREL/TP-580-26157, 72 pp.

Yu, E.K.C., Deschatelets, L., Saddler, J.N., 1984. The bioconversion of wood hydrolyzates to butanol and butanediol. Biotechnol. Lett. 6 (5), 327–332.

Zverlov, V.V., Berezina, O., et al., 2006. Bacterial acetone and butanol production by industrial fermentation in the Soviet Union: use of hydrolyzed agricultural waste for biorefinery. Appl. Microbiol. Biotechnol. 71, 587–597.

Chapter 3

Political Orientations, State Regulation and Biofuels in the Context of the Food–Energy–Climate Change Trilemma

M. Harvey[1] and Z.P. Bharucha[2]
[1]*Centre for Economic Sociology and Innovation, Department of Sociology, University of Essex, Colchester, United Kingdom* [2]*Department of Sociology, University of Essex, Colchester, United Kingdom*

INTRODUCTION

This chapter will explore the importance of different political orientations in developing regulatory frameworks for biofuels in the context of competing or complementary uses for land resources, food and energy demand, and their significance for climate change mitigation. The politics of both energy and food security are intimately connected to those of climate change mitigation, but in contrasting ways in different national contexts. National strategies for both energy and food security are strongly conditioned by the spatial location of resources *either* of land, sun and water for food and biomass for energy; *or* of fossil fuels, conventional and nonconventional. In the short run, turbulence in fossil energy markets, especially oil, is the likely prelude to the longer-term necessity for replacing finite with renewable energy resources. Shaped by this turbulence, different renewable alternatives to fossil carbon energy present themselves very differently in different environmental resource contexts. Based on on-going research, this chapter will deploy a neo-Polanyian approach in which biofuel production and markets are discussed as "politically instituted" through the use of mandates and dedicated sustainability regulation (Harvey, 2014; Harvey and Pilgrim, 2011, 2013). In that respect biofuels are contrasted with food production, the other and much more significant driver for land-use change, with correspondingly greater environmental impacts (including as a source of greenhouse gas (GHG) emissions). Indeed, in some national contexts, the priority of food production is politically constructed as the logic behind biofuel regulation. Finally, political strategies towards energy security and reliance on fossil carbon energy condition national strategies for climate change with respect to both a low-carbon sustainable intensification of agriculture and renewable biomass energy. In order to substantiate this argument, the chapter will compare recent developments in political strategies and regulatory regimes towards the challenges of the food–energy–climate change trilemma (hereafter "trilemma") in Europe (with a particular focus on Germany), China, Brazil and India. The chapter aims to develop the debate over the politics and regulation of biofuels (Borras et al., 2011) by placing it within the broader political context of the trilemma challenge as it has developed in recent years.

The historical development of biofuel markets has long been distinguished by the decisive role of the state in politically shaping their trajectories, whether in Brazil, the United States or Europe. This is hardly surprising, given that energy security has been and remains a national strategic priority. Thus, the first "take-off" of biofuel production and market formation was powerfully stimulated by the threats to Middle East supplies and the oil-price shocks of the 1970s, notably in Brazil with its ProAlcool program for ethanol from sugarcane, and in the United States for ethanol from maize (Harvey and McMeekin, 2010; Gee and McMeekin, 2011). The neo-Polanyian concept of "politically instituted markets" captures the significance of the role of the state first by emphasizing the need to comprehend how the organization of economies is instituted in space and time, and then, additionally to analyze how, *in particular historical circumstances*, states and political agencies provide the conditions of emergence of a novel economic organization of production, distribution and markets. Extreme examples of politically instituted markets, highlighted by Polanyi (2015), were the military command economies of the Second World War, which were profoundly, and in some cases, enduringly transformative. More contemporary examples include the privatizations of previously public organizations, where the nature of, and relations between, service providers, infrastructure developers, pricing and product-quality regulators are defined and legally established by governments. Even the bastions of free market ideology can conceal the characteristics of a strong state-directive role more typical of developing economies when inducing major technological transitions (Block, 2007, 2008). The concept of "politically instituting" includes the role of the state in regulating and legalizing markets, but radically extends that role in order to encompass historical examples of where the state itself engenders the emergence of novel economic organization, often with novel products and markets. Biofuels, it is argued, are a case in point.

Just as the initial take-off of biofuel transport energy economies was stimulated by issues of energy security and development, their subsequent nationally diverse developments have been shaped by complex internal and external conditions. Since the 1970s, recognition of climate change driven by GHGs from societally varied sources (eg, power generation, transport, domestic consumption, agriculture, land conversion) has emerged as a major political context shaping biofuel markets and their regulation. In the wake of the spikes in both oil and food prices immediately preceding the 2007–08 financial crisis, the perceived new competition for land between production for food and energy transformed the political context of biofuels, but in starkly contrasting ways in Brazil, Europe and the United States (Smith et al., 2010; Harvey and Pilgrim, 2013). Thereafter, the rapid growth of nonconventional oil and gas supplies, notably from fracking, combined with the global economic slowdown and the dramatic fall in oil prices altered the global and geopolitical landscape of energy security, especially in the diminished dependency of the United States on Middle East oil. Whether in terms of market competitiveness or energy security, these developments have altered the political context for developing biofuel economies. Moreover, political contexts are always arenas for political controversies: biofuels are seen as a renewable source of energy mitigating climate change by some, or a "cure worse than the disease" by others (Doornbosch and Steenblick, 2007); Republicans in the United States are not the only powerful political influences denying the role of human activity in causing climate change; following a moratorium in 2011, Germany appears set to ban fracking across much of its territory, while the United States vigorously promotes it; some countries ban the use of food crops or even land producing food crops for the production of biofuels. The conflicts and power dynamics between different political forces form an important part of analyzing the processes by which biofuel markets are politically instituted.

THE TRILEMMA APPROACH

The concept of the food–energy–climate change trilemma highlights some key dynamics behind the political institution of biofuel markets with its emphasis on the connections between the global drive to produce more food and more energy for growing and developing populations and economies, under the constraints of climate change and finite planetary resources of land, water and fossil carbon energy. However, these connections have exhibited major geopolitical shifts, with varying consequences in different regions—but, at least in the short term—mostly negative in terms of biofuel markets. Different political economies, with different political systems and orientations, confront very contrasting access to the planet's environmental and finite resources (notably land and water; fossil energy whether conventional or nonconventional oil and gas; or sun). In the first decade of the 21st century, with rapidly growing demand, there was much discussion of "peak oil," and the volatility of oil prices, including dire predictions by the IEA, and scientific papers attesting that peak oil production had possibly already been passed. Indeed, in terms of dire consequences, high oil prices were intimately connected to the collapse of the subprime mortgage bubble, triggering the financial crisis (Taylor, 2009). In this context, biofuels appeared to offer a win–win potential, both in contributing to alternatives to diminishing resources of a finite oil resource and with climate-change mitigating benefits. Postcrash, from the standpoint of energy security and finitude of oil resources, if not for climate change consequences of burning fossil fuels, the whole scenario changed. It has long been realized that "peak oil" was always as much a political as a straightforward physical resource issue, where political control and geographical location of fossil carbon energy by states and oil majors, were as much at the root of market turbulence as the swings of supply and demand. The current phase of turbulence is thus marked by the clash between the US drive for greater energy self-sufficiency and OPEC's, and in particular Saudi Arabia's, strategy of driving down oil prices. From 2010 to 2014, US oil production increased by almost 60% from 5.5 to 8.7 million barrels per day (mmbd), compared with Saudi Arabia's 8.9 to 9.7 mmbd (US Energy Information Administration, 2015). From Nov. 2014, Saudi Arabia as the leading player in OPEC promoted a new strategic policy of maintaining oil supplies at a high level with the aim of perpetuating its dominant market share. The policy is equally directed at undermining investment in relatively high-cost, high-risk oil exploration and oilfield development, whether deep ocean or in nonconventional fracking. Yet, of course, "peak oil" has not gone away. Resources *are*, ultimately, finite. The IEA's prediction is for the United States to fall back to 1990 levels of production by 2035 (IEA World Energy Outlook, 2012). Perhaps more remarkably, Ali al-Naimi, the kingdom's oil minister told a conference in Paris on business and climate change: "In Saudi Arabia, we recognise that eventually, one of these days, we are not going to need fossil fuels. I don't know when, in 2040, 2050 or thereafter" (*Financial Times*, May 21, 2015). Yet, the current phase of turbulence in conventional oil markets places all competitors to cheap oil, including biofuels, in a hostile market environment.

Of possibly equal significance has been a geopolitical shift in demand for land for food. Following China's inclusion in the WTO from the late 1990s, there has been a major restructuring of world markets for key agricultural commodities, particularly focused on Brazil as a major exporter of red meat, poultry, soybean and maize. Significantly altering its previous policy of food self-sufficiency, China has become a major new source of international demand for food, exemplified by its bilateral agreement with Brazil for the import of soy for animal feed (Sharma, 2014), introducing a new global link between trilemma challenges in Brazil and China (Bharucha, 2014). Consequently, there is now additional competition for land and for potential expansion of agricultural land into Amazonian and Cerrado biomes in Brazil. Within Brazil, pressure on land extensification has in turn stimulated the emergence of a range of new regulatory and self-regulatory measures, which have demonstrated varying levels of effectiveness, including the Zero Deforestation Policy and the Cadastral Land Registry, the Roundtable for Responsible Soy and the Soy Moratorium (Gibbs et al., 2015). The regulatory context for developing a biodiesel industry from soy is thus being modified, in part at least because of the joint competition for land for food and fuel and in the context of new transnational supply chains. The spread of new varieties of sugarcane for bioethanol into the Matto Grasso is a further indication of the changing shape of competition for land, and its regulatory environment. However, the conjuncture between high oil prices and high food prices in 2007–08, prior to the great crash, equally prompted major regulatory shifts both in Europe and China. China effectively proscribed the use of crops or croplands dedicated to food production for the development of its own national biofuel industry, largely closing off this particular innovation pathway. At the same time, as we shall see below, European regulation equally developed a policy to reduce or eliminate competition between food and biofuels for either crops or land. Viewed from a neo-Polanyian perspective, these shifts are indicators of how both competition and insulation from competition for land between food and bioenergy production has been politically instituted, variously across different regions of the world. They are politics generated by the trilemma as it manifests itself in different regions, for different political economies with access to different resource environments (namely fossil fuels, land, sun and water).

To explore the dynamics of different emergent regulatory frameworks in different regions, we now turn to a comparison between Europe (Germany), India, China and Brazil. The aim is not to present comprehensive case studies, but to focus on the key dimensions of regulatory variation in each country.

BIOFUELS IN THE EU AND GERMANY

Trilemma challenges have a very distinctive expression in Europe in general, and Germany in particular, which have profoundly shaped European regulatory strategies towards biofuels. The European regulatory response provides the framing context, within which different nations have pursued their own policy variations on the European theme (Harvey and Pilgrim, 2013). As a broad characterization, to be explored below, the context for European policy and regulation for biofuels has overwhelmingly been climate change mitigation, with energy security as a far less dominant driver, in spite of the high dependency on imports, and the uncertainties and risks stemming from the conflict with Russia, especially surrounding Ukraine in recent years. Similarly food security, and the 2007–08 food price spikes as such, played only a limited and indirect role in forming biofuel promotion, demotion or regulation in Europe. Land, and land use, features in a uniquely European fashion in trilemma terms. The puzzle to be examined here, therefore, is why transport energy has slipped from being a center for climate policy attention to become parked in the long grass, in Europe as a whole, and distinctively so in Germany.

The puzzle is deepened by the European realities of GHG emissions. Though emissions from European aviation and shipping demonstrate the fastest increases, road transport is a significant emitter, contributing around one-fifth of the EU's total CO_2 emissions and responsible for just over 71% of all transport-related emissions (European Commission, 2015a, citing 2012 data). Emission reductions from transport have not kept pace with those achieved in other sectors. Between 1990 and 2007, transport emissions rose by 29%, while emissions from other sectors fell by 24% over the same period. A short dip in emissions between 2007 and 2009 was primarily attributed to the global recession, with evidence of a resumed upward curve as Europe recovered from 2010 onwards (Hill et al., 2012). These figures underscore political urgency of dealing with emissions in transport, as without declines in this sector the achievement of EU-wide targets is unlikely (Hill et al., 2012). The scale of the challenge is significant. While the EU White Paper on Transport (European Commission, 2011) calls for a 60% reduction in transport-related GHGs by 2050 relative to 1990 levels, in real terms, a 67% reduction must be achieved by 2050 to meet the long-term GHG reduction target set out in the White Paper (European Environment Agency, 2015).

Biofuels in Europe

In sharp contrast with Brazil and the United States, where biofuel markets were driven overwhelmingly by the imperatives of energy security from the 1970s onward, Europe first began a significant program promoting biofuels decades later, and primarily as a response to the Kyoto Protocol, as a means of decarbonizing transport (Pilgrim and Harvey, 2010). The 2003 Renewable Energy Directive (2003/30/EC) set the first targets for biofuel consumption in member states. These nonbinding targets were replaced by two Directives issued in 2009, which have come to govern the deployment of biofuels across Europe. These Directives already mark a major retreat from direct promotion of biofuels, resulting from controversies surrounding biofuels discussed below. The Fuel Quality Directive (FQD) sets climate mitigation criteria for *any* fuels used in transport, stipulating a 6% reduction in GHG emissions from each unit of energy supplied by 2020. The Renewable Energy Directive (RED) mandates that renewables should contribute 10% of the final consumption of transport energy in each member state by 2020, but biofuels were no longer specified or required.

The premise that biofuels would represent a GHG-saving relative to fossil fuels was thus fundamental to the project, and clarifying this premise has been a core preoccupation occupying all actors in the market. For biofuels to qualify towards the GHG-reduction targets under the 2009 RED, they must achieve a 35% GHG-saving relative to fossil fuels, rising to 50% in 2017 and 60% in 2018. With the FQD, the overall regulatory framework locks in the imperative that transport energy overall must demonstrate measureable GHG savings, but retreats from mandates or direct promotion, especially of so-called first-generation biofuels.

The contribution of foodcrop-based biofuels to 2020 targets, and specifically the question of whether they represent a genuine GHG-saving, has been the subject of tremendous controversy which has uniquely shaped the structure, size and ambition of European biofuel markets. Concerns mainly revolve around the complex interactions between biofuels policy, GHG targets and the impacts of land-use change. The conversion of noncultivated land for biofuel production (direct land-use change, dLUC), or the displacement of pre-existing agricultural activity onto noncultivated landscapes (indirect land-use change, ILUC) raised concerns that biofuel policy could introduce a variety of perverse impacts. Two categories of impact have particularly preoccupied discourse and policy—assumed negative correlations between global food prices and the use of food crops for fuel (the "food versus fuel" debate), and more recently, the awareness that land-use change driven by biofuels policy introduces an additional source of GHG emissions which negate any potential savings which might derive from the substitution of fossil fuels with biofuels.

dLUC is effectively mitigated by existing policies—sustainability criteria govern which biofuels can count towards emissions reduction targets, and these effectively prevent the direct conversion of noncultivated lands for feedstock cultivation. By contrast, the construction of an ILUC policy has been rather more contentious, in part stemming from the essentially "irresolvable uncertainties that ultimately characterise knowledge about the geographical footprint of biofuel-driven land-use change" (Palmer, 2014, p. 342).

The adoption of the 2009 RED and revised FQD coincided with scholarly debates and environmentalist campaigning around the Searchinger and Fargione papers warning that ILUC-generated GHG emissions potentially implied even worse climate impacts than those generated by conventional fossil fuels. Subsequently, the EU agreed to initiate a process of additional research to clarify the issue, and revise targets accordingly. What followed was a protracted and contentious debate about the "ILUC file," "one of the most debated" files to circulate through the EU Parliament (Eickhout, 2015).

Concerns about the carbon implications of EU biofuels are shaped by distinctive interactions between food and fuel feedstocks, related to the dominance of diesel in the European transport fleet. This has particular implications for the nature of the debate around biofuels and land-use change in Europe. The dominance of biodiesel in total biofuel mix limits the impact of EU biofuel consumption on food grain consumption (European Commission, 2015b). Instead, concerns revolve around competition between food and fuel uses for rapeseed, and the alleviation of these by the import of southeast Asian palm oil. In other words, the key concern is that increased use of rapeseed for biodiesel diverts it from its uses in food and industry, and thus "pulls in" palm oil—the most cheaply available substitute. Resulting forest clearances raise the specter of increased GHG emissions, deepening concerns that "Europe's climate policies are creating climate change instead of reducing it" (Eickhout, 2015). So, what is distinctive about the whole European ILUC controversy, and its impact on regulatory frameworks for biofuels, is that it concerns *land outside Europe*. Only recently have GHG emissions from European industrialized agriculture entered the policy arena, and even the Common Agricultural Policy (CAP) reforms of the late 1990s subsidizing the setting-aside of cultivated land was directed primarily at reducing the overproduction resulting from production subsidies and protecting and increasing biodiversity (Krebs et al., 1999; Piorr et al., 2009), rather than climate change mitigation. There is thus a marked contrast between the biofuel (agrofuel) furore over ILUC outside European territories (in Indonesia, Brazil, etc.), and the relative silence over GHG emissions from EU food production.

In the absence of achieving any robust and universally agreed assessment of the extent and precise implications of ILUC, controversy flourished. Finally, in Jul. 2015, the EU Parliament froze the issue by voting in an ILUC Directive which caps at 7% the contribution of foodcrop-based biofuels to the RED target of 10% renewables in EU transport energy. This development allows some room for the further expansion of conventional biofuels in the EU transport mix (currently renewables contribute only some 5.7% of the EU's total transport energy), but imposes a definite upper limit on their contribution, and signals strong policy support for the development of advanced substitutes. Given the dominance of biodiesel in European biofuel consumption, and the negligible innovation in 2G biodiesel, Europe has now created a regulatory context inhibiting the development of biofuels as a response to climate change mitigation. Meanwhile, emissions from transport energy continue to rise.

We're talking less than 4.7% of our final EU transport energy consumption coming from biofuels. We're having a big argument about whether they're the wrong biofuels, but we're kind of missing the 95% picture which is that it's oil coming out of the ground somewhere. Now of course that's fine if you're saying that the biofuels you're using are actually worse than oil, which they may be in some senses. But you've got to recognize that bigger picture and work out what our real ambition is…

Interview with expert on European environmental policy, Nov. 11, 2014

And the length and fractious nature of the debate has had a significant impact on those engaged with biofuel markets, shaking confidence in the political backing which supports them.

The whole discussion over the last 3 years [is such that] as an investor I would say no, I'm not stupid, I wouldn't get into this. We don't have any goals after 2020, the new 2030 have nothing about alternative fuels and transportation. There is no movement for investment.

Interview with representative of international NGO, Nov. 27, 2014; parentheses added.

Germany

Within this Europe-wide regulatory environment, Germany, as the largest EU economy and consumer of fossil fuel for transport, developed a particular trajectory of first promotion and then restriction of biofuel development over a brief timespan of a decade and a half. Germany already had a small-scale history of renewables in transport predating the implementation of EU targets and regulations, beginning with a modest "1000 tractor program" in the 1990s. Then, in the early 2000s following the 2003 EU directive, tax incentives supported the rapid development of a domestic market and expansion of biofuel plants which sold directly to agriculture or truck fleets. Biofuels, and particularly 100% biodiesel fuel, were exempt from taxation in the Mineral Oil Duty Act (2004), which was then at €0.67 on petrol and €0.47 on fossil fuels, giving biofuels a competitive advantage. This industry wove together rural industry and agriculture. Agricultural associations began to develop capacities for storage and distribution and a great deal of domestic expertise was developed. Confidence and enthusiasm was high:

> ... *that was really good business, and everybody was building biofuels plants in say 2004, 2005, 2006. It was like someone told me it was a time like in the US with gold-digging. It was so easy to make money.*
> Interview with representative of German biofuel industry, Nov. 28, 2014.

Then came a significant shift, whereby major oil companies became responsible for selling blended fuels. Domestic biofuel producers could no longer compete with larger players whose requirements were several orders of magnitude higher than they could manage:

> ... *most of the small biofuels producers had no chance on that market because the mineral oil companies said yeah, I don't want to buy 1000 tonnes, I need to have 100,000 tonnes. So it was a market where only the big biofuel producers had a chance. And you can see it in Germany at the moment, there are just the big producers left at the market. Most of the small ones disappeared.*
> Interview with representative of German biofuel industry, Nov. 28, 2014.

Whereas domestically grown rapeseed, and some residues from potato and sugar beet, had been the dominant feedstocks, *"(with) the mineral oil companies coming in the feedstock sources changed. It became about what was the cheapest feedstock, depending on the market prices"* (Interview with representative of EU environmental NGO, Nov. 27, 2014). Rapid growth in domestic consumption from 450,000 tonnes in 2004 peaked at 3.26 million tonnes in 2007, with the cost in taxation exemptions rising from €520 million in 2004 to €2144 million in 2007 (Rauch and Thöne, 2012).

However, this boom period for biofuel growth, and particularly biodiesel, was short-lived, brought to a close both by the 2008–09 shift in EU policy discussed above, and several additional drivers, including environmentalist concern over ILUC and monocrop agriculture; the policy reorientation to renewables for power generation under the "energy turn" (Energiewende); and the particularly German cultural love affair with high-speed automobility on the autobahn network. Responding to the same controversies that had affected EU-wide regulatory policy, major changes in the regulatory framework were introduced in Germany, resulting in the confinement of biofuels to a small fraction of transport energy and the reversal of a growth strategy. Firstly, from 2008 tax exemptions were progressively withdrawn, and already by 2010 the cost of the exemption reduced from its peak of €2144 million in 2007 to a mere €10 million in 2010. Secondly, the 2007 Biofuel Quota Act effectively replaced tax exemptions with mandated quotas for biodiesel blending set at 4.4%, well below the share of biofuels in the transport fuel mix in 2007 (5.7%) (Rauch and Thöne, 2012). The impact of these regulatory changes was rapid and twofold: the volume of biodiesel consumed fell by a third within 6 years, from its peak of 3.26 million tonnes, down to 2.16 million tonnes in 2013; and the share of pure biodiesel compared with blended diesel dropped from 56% in 2007 to a mere 1% in 2013 (UFOP, 2013a,b). It signaled the death of the 100% biofuel vehicle in Europe, as a present or future prospect.

This quota system was *further* replaced in 2015 with a GHG-reduction commitment, whereby, in compliance with EU requirements, mineral oil sellers need to reduce the GHG emissions of transport fuels by 3.5% (rising to 4% in 2017 and 6% in 2020). The general rationale behind this shift was welcomed by some environmentalist lobbies: "*it's good to see that they're focussing on the end result which we want to achieve... That seems to be a much better way to focus. We want to reduce GHGs, so let's aim to do that*" (Interview with researcher on European environmental policy, Nov. 11, 2014). But for established actors in the market, the switch introduced a great deal of uncertainty around continued political support for conventional biofuels. The implementation of the new regime:

> *... introduces a completely new system, (with) completely new pricing... that again the companies need to adapt to and understand first... How to sell a completely new product in the end, because you've never (before) sold something based on greenhouse gas reduction... So that also in total... that's something that really difficult, really challenging for biofuels producer to stay in the market.*
>
> Interview with representative of German biofuel industry,
> Nov. 28, 2014, parentheses added.

It has also been pointed out that the new system paradoxically results in *lower* volumes of biofuels used—as the greater the GHG-saving achieved by a biofuel, the *less* of it is required in the final product.

Apart from the Europe-wide obsession with ILUC, and the shaping of Germany's biofuels market by the application of EU requirements (notably the recent switch to the GHG-reduction commitment), Germany has been distinctively characterized by the lack of *popular support* for conventional land-based feedstocks stemming from concerns about land-use change. The feedstock for German biodiesel was overwhelmingly from rapeseed grown domestically, spreading the "sea of yellow" visible across Europe, with Germany leading the way. While the total amount of palm oil in the EU's biodiesel consumption is limited, the diversion of rapeseed oil to biodiesel from other uses sucked in palm oil to replace it for food and industrial uses, provoking debates about biodiesel driving tropical deforestation outside Europe:

> *I think what the German public realized is that we don't have the potential here. We were importing palm oil from Malaysia for instance. And people were increasingly uncomfortable with that, and the question became shouldn't we be doing something here and not importing?*

Germany can't replace the amount of oil we consume without getting something from outside. So the issue is, do we want to shift our problems to Malaysia, to Mexico, and the German public has essentially decided, no.
Interview with expert on the German energy transition, Nov. 24, 2014.

These concerns and tensions have acted to isolate the project of transitions away from fossil fuels in transport from the *Energiewende* underway in Germany. This broader movement emerged from the wave of popular community-level protests against nuclear power that developed from the 1970s onwards— "*the most powerful social movement in Germany*" (Interview with expert on German social movements, Nov. 25, 2014). These protests catalyzed a social conversation about renewables particularly for the electricity sector, and a wave of community-driven wind and biomass power projects. These were subsequently given political support via the Renewable Energy Act (2000, amended 2004, 2006), focusing on electrical power generation (Langniss et al., 2009). Community-led renewables for electricity have since remained robust, supported by Germany's strong federalism and a robust "*politics of local communities... citizens sometimes coming together and forcing the hand of the municipality*" (Interview with expert on the German energy transition, Nov. 24, 2014). Over time, the *Energiewende* or energy transition has evolved into a major national project that enjoys cross-party support, builds on the dynamism and political action of a popular social movement and has succeeded in instituting a qualitatively different regime in the power sector (Strunz, 2014). Progress towards alternative fuels for transport has not "joined the parade." In the absence of a popular social movement pushing for alternatives in transport and with biofuels stigmatized by complicated and contentious sustainability credentials, transport energy has effectively been sidelined within the *Energiewende*.

Moreover, green electricity power generation does not translate into electric cars for renewable transport energy. The German automobility culture of high-speed cars for long-distance travel to get them to the beach first in front of all other Europeans, is reflected in the derisory 25,000 electric vehicle ownership in 2015, a tiny fraction of the 3 million car registrations, less than half the take-up of France, and two-thirds of the take-up in the United Kingdom. A study of consumer preferences indicated a much higher demand for high-range, high-performance vehicles, with a maximum predicted adoption of electric vehicles of 5% often as second cars, matching government expectations (Lieven et al., 2011). Yet there is still some hope for change:

Germany has always been a country where people identify with their cars, it's a status symbol. We are arguably the most car crazy country in the world. But this is changing. Young people are happy not to own a car, they're happy to take public transportation and have the best smartphone, navigating their way around. Cars are losing their status symbol.
Interview with expert on the German energy transition, Nov. 24, 2014.

In sum, at this point in time, Germany is distinctively responding to trilemma challenges, advancing in renewable energy overall, but locked in to fossil fuel transport energy. The ILUC controversy, with its particular resonance in Germany; the environmentalist focus on power generation; and German car culture, have combined to block the early trajectory of biofuels without providing any significant alternative route out of high-fossil carbon dependency for road transport.

INDIA: FOOD VERSUS FUEL AND STILLBORN BIOFUEL MARKETS

As a developing economy, India's trilemma and its manifestation in biofuels policy, naturally contrasts with an advanced economy such as Germany, but is equally distinctive when compared with China and Brazil, as we shall see. The puzzle for India can be expressed in terms of global sugar league tables: up until the mid-1970s India was the number one producer of sugar in the world, followed by Cuba, and then Brazil. Brazil then took the top spot in sugar production, overwhelmingly as a consequence of its bioethanol from sugarcane program, primarily in response to energy security and the oil-price shocks of the 1970s. Although, as we shall see, India was even more dependent on imported oil than Brazil, it did not respond in like fashion. India has since retained a number two sugarcane production position, but the gap with Brazil in terms of production and hectarage progressively widened, as India remained hooked on the sweetness of sugar, while Brazilians drove around in biofueled cars (Fischer et al., 2008). Between 2002 and 2013 while Brazil more than doubled its hectarage and production of sugarcane, India witnessed modest increases of 10–15%. Nonetheless, India still produces nearly three times more than the number three producer in the sugarcane league table—China (FAOSTAT, 2015).

So, in the following analysis we will seek to explain why there was no biofuel revolution in India. Following sporadic and limited efforts from the 1970s onwards, the Indian government first outlined concrete mandates for biofuels in 2001–02, trialing a 5% ethanol blend (E5) in nine states and four union territories. Also in the early 2000s, a National Mission on Biodiesel was set up, which aimed to produce biodiesel from *Jatropha curcas* on so-called wastelands (noncultivated rural commons), to enable 20% blending with high-speed diesel by 2012. A national Biofuels Policy was only finally formulated in 2008. This set out "indicative" targets for 20% blends for both ethanol and biodiesel by 2020—relatively ambitious in global terms. Yet, in 2012, India accounted for just 1% of global biofuel production. Blending targets *"have never looked like being reached,"* particularly for biodiesel, where production is so low that blended fuel *"never gets into the organized transport sector"* (Interview with experts working on advanced biofuels in India, Jan. 2015). This has emerged as a result of a uniquely Indian configuration "pulling" land-use away from biofuels and "pushing" food production ahead of transitions away from conventional fuels.

We don't see the feasibility of doing what other countries have done, which is why the uptake of biofuels in India has not been that great even though transport is a huge sector in India with a huge need.

Interview with a climate change NGO, Jan. 2015.

Though India is currently the world's fourth largest energy consumer, and the pace of growing demand is amongst the fastest in the world (Luthra, 2014), nearly a quarter of Indians lack access to electricity (Ahn and Graczyk, 2012) and some 700 million are still directly reliant on biomass for household cooking and heating (Planning Commission, Government of India, 2014). Transitions in transport are similarly conditioned by expected increases in energy demand. At 12 passenger cars per 1000 people in 2009, Indian car ownership is lower than that of China (34 cars per 1000 people). Car ownership is estimated to rise to around 100 per 1000 people in 2035 (IEA World Energy Outlook, 2014). Together, expected increases in demand across different sectors mean that even allowing for a reduction in energy intensity (eg, through increased efficiency), India's primary energy supply will need to grow by around 5.8% annually (Planning Commission, Government of India, 2014).

As is the case with Europe, urgent energy security concerns have not driven attempts to become less dependent on fossil fuels for transport energy. India is the fourth largest consumer of oil globally and the fourth largest importer (IEA, 2012). Domestic hydrocarbon reserves are negligible—just 0.3% of global reserves. There is a high degree of dependence on imported fossil fuels, particularly in transport. Ninety-five percent of India's transport energy is derived from petroleum and just over three-fourths of crude oil consumed is imported. Crude oil imports have risen substantially—from 21 Mtoe imported in 1990 to 162 Mtoe in 2009 (Ahn and Graczyk, 2012). In light of expected increases in transport energy consumption, securing or "topping up" energy supplies is a key driver of Indian energy policy, with a continued commitment to fossil fuel sources either imported or domestically sourced. In Aug. 2015 India's Prime Minister Narendra Modi visited the UAE, with driving forward oil trade an important policy objective for the visit (Roy, 2015).

Carbon emissions from the consumption of fossil fuels make up a very significant 54% (year 2000) of India's total GHG emissions (Planning Commission, Government of India, 2014). India has a goal of reducing GHG emissions per unit of GDP by up to 25% below 2005 levels by 2020. Relative to the European Union, however, decarbonization has been less of a driver of energy transition than energy security. Per capita transport emissions are still relatively low (ICCT, 2015), and threats to public health from vehicular pollution have been much more prominently problematized than the need to mitigate GHG emissions. Particularly in urban areas, a rapidly growing vehicle fleet combined with poor road infrastructure and inadequate fuels and vehicle standards has driven a substantial spike in air pollution. It is estimated that transport emissions caused some 49,500 deaths in 2010, and under a business-as-usual scenario, it is estimated that these could rise to around 158,500 by 2030 (UrbanEmissions.info, 2012). In early 2015 a visit to New Delhi by US President Obama foregrounded India's problems with air pollution, when the US Embassy reportedly ordered some 1800 Swedish air purifiers and officials expressed concern about Obama being outdoors to attend India's Republic Day parade, where he was invited as Chief Guest. Since then, a number of media and civil society campaigns have stepped up pressure on the government to address urban air quality, and have foregrounded the importance of better fuel quality standards and improved transport infrastructure.

Climate policy per se, by contrast, has been described as *"very ad hoc and unstructured so far,"* with policy instruments acting *"more like guiding documents and guidelines [rather than] clearcut directions on what we should be doing"* (Interview with representative of a national NGO working on climate change and the environment, Jan. 2015). Interactions between climate policy and renewable energy policy have been weak at best. India's National Action Plan on Climate Change, formulated in 2008, calls for a 1% annual increase in renewable energy capacity as part of a GHG-reduction strategy, but the dominant stance is of continued commitment to fossil fuels to drive development. Commitments to replace fossil fuels with renewable sources or to cap emissions by a certain date are considered impractical in light of the perceived urgency of food and energy security:

India needs to develop more to ensure assured access to food, water, etc. If China, which is in many fronts 15–20 years ahead of India in the development curve, will peak in 2030, then going by that logic, India cannot peak before that it and it will be at least 5–10 years after that.

Interview with representative of a national NGO working on climate change and the environment, Jan. 2015.

And:

India is not too obsessed with the carbon emissions angle. They [policymakers] are getting more and more concerned from an energy security perspective. I don't think there is any intention to bring down the share of fossil fuels. India is mostly concerned with energy security. They would be happy to explore more and more for conventional fuels and I don't think that is going to stop.

Interview with biofuels scientist, Jan. 2015.

This is not to say that GHG savings and climate change mitigation do not appear on the transport policy agenda at all. The fact of climate change and the imperative to mitigate rising emissions are both mentioned in passing in biofuels policy documents. However, it is clear that no significant policy action towards biofuels has been driven by either climate change or energy security imperatives.

Indian Biofuels Policy and the Trilemma

Policy rhetoric has invoked strong "win–win" narratives (Pradhan and Ruysenaar, 2014) around the potential for biofuels to rejuvenate rural and agricultural livelihoods, "[generating] millions of additional on-farm jobs and lucrative markets for farm produce" (Planning Commission, Government of India, 2002, p. 5). However, the reality is that given the overriding priority accorded to food production, and the multiple obstacles to agricultural reform relating to land-holding, caste and rural poverty, the barriers to meeting the proclaimed targets appear insurmountable. Perhaps the strongest influence on the emerging biofuels sector has been an early commitment to avoid a food-versus-fuel tradeoff by preventing the diversion of arable land for the cultivation of energy crops.

India supports a population of some 1.1 billion people on around 2.4% of the world's land and with 4% of global water resources. Some 194.6 million people are undernourished in India (FAO, 2015), equivalent to around 15% of the population. Agricultural intensification alone will not address this challenge, but chronic agrarian crises will not help either. Early efforts at intensification, notably the Green Revolution during the second half of the 20th century resulted in substantial increases in cereal output, but largely failed to address the complex problems of hunger and poverty.

Increased yields in the intensively irrigated landscapes of the Green Revolution preceded significant sustainability crises. Free electricity to facilitate groundwater pumping and the construction of huge irrigation infrastructures have generated a number of social–ecological crises, including falling aquifer levels (Rodell et al., 2009; Tiwari et al., 2009), the displacement of some 40 million people by dam and canal projects (D'Souza, 2008) and irrigation-induced soil salinization over some million hectares of prime agricultural land in northwestern India (Datta and de Jong, 2002). At the same time, the relative neglect of rainfed landscapes, which constitute between 60% and 70% of India's agricultural land, has entrenched land degradation. Poverty and hunger loosely overlap, with some 30% of the population in India's degraded semi-arid watersheds living below the poverty line (Ryan and Spencer, 2001). To these longstanding crises of agriculture and hunger are added new pressures from climate change on the one hand, and a growing and increasingly affluent population on the other. Changing food demand will require a 50% increase in food production by 2030, while climate change is expected to severely affect yields, particularly in nonirrigated landscapes which have so far depended on the remarkably stable seasonal monsoon rainfall (Turner, 2013).

Pressures on agricultural land, then, are immense and urgent, requiring new forms of sustainable intensification (Pretty and Bharucha, 2014) that do not replicate the unsustainable outcomes generated by the Green Revolution. For some actors involved with agricultural and rural development, the urgency of the challenge is not matched by the speed or scope or effectiveness of the policy response: "*Agriculture is a very confused sector in India and has been a neglected sector for ages. We have not really made up our minds about what to recommend*" (Interview with expert on Indian agriculture and energy policy, Jan. 2015). Implementation of existing agricultural policies is considered ineffectual: "*Government policies in the agrarian set-up are namesake policies. They cannot be implemented, they cannot be enforced, they cannot be monitored, they cannot be assessed"* (Interview with civil society leader, Jan. 2015). In this context then, it seems unlikely that sustainable intensification will be able to effectively alleviate pressures on land and contribute to food and energy security.

Bioethanol

Bioethanol made up an estimated 2.1% of India's total fuel market in 2014, expected to rise to around 2.5% in 2015 (USDA, 2014). Sugarcane is the dominant feedstock, with molasses derived from sugar production used to produce ethanol. Of this ethanol produced, most is acquired for alcohol-based chemical manufacturing (40%) and for the production of potable alcohol (45%) with only a small percentage remaining available for other uses, including as a fuel-blend (Ray et al., 2011).

Despite well-established production and supply chains for ethanol, the procurement of adequate ethanol to meet targets has been problematic for a number of reasons. Sugarcane availability in India follows a cyclical pattern; production is affected by prices and water availability and ethanol targets have had to be scaled back in the early 2000s to accommodate shortfalls in availability.

More recently, falling oil prices and competition from industrial applications and potable alcohol have made it unviable for India's oil-marketing companies (OMCs) to procure ethanol for blending. In late 2014, for example, India's Ministry of Petroleum and Natural Gas announced that OMCs were prohibited from procuring ethanol from sugar refineries unless their offer prices were reduced to take into account the fall in global oil prices. Refineries in turn declined to change procurement prices until the 2015–16 fiscal year (ICIS, 2014). For refineries, selling ethanol for industrial alcohol was more profitable and, as a result, tenders for the procurement of some 1200 million liters of ethanol were canceled by the OMCs (Jha, 2014). Thus, strong competition from other, established industries and the lack of a strong policy response enabling the diversion of ethanol to biofuels has resulted in lackluster progress towards blending targets. In the absence of this support, neither farmers nor refineries are particularly keen to increase the availability of sugarcane nor to prioritize ethanol production for the biofuel sector: *"Ethanol policy hasn't done anything to the sugarcane crop, because everybody still thinks in terms of sugar"* (Interview with civil society leader, Jan. 2015). There is also considerable resistance to reforming sugarcane economies because of the nature of entrenched interests in the industry. Even where sugarcane has acquired a privileged position in an agricultural region, the capacity to develop a biofuel industry has been blocked, exemplified by the case of Maharashtra state in peninsular India. Here, sugarcane is a "political" crop, *"a symbol of political power [which] has nothing to do with profitability... Nearly everything in a sugar factory is subsidized by the state"* (Interview with expert on agricultural sustainability and water management, Jan. 2015). The political corruption of the industry has resulted in the diversion of water for sugarcane cultivation and processing at the expense of other crops, including those important for local food security. Some 60% of Maharashtra's irrigation water is appropriated by sugarcane planted on under 5% of cultivated land. This has driven the relative marginalization of other crops necessary for food security and better adapted to the semiarid growing conditions of peninsular India:

> *Oilseeds... require less water and can grow in arid zones. This year [2015, in Nagpur, Maharashtra], of the 16,000 ha expected of oilseed [cultivation], only 2000 ha are actually cultivated. Many oilseeds have just not been planted. Instead, they are cultivating sugarcane because of the sort of security that the sugar factories give them. There is no such security when it comes to oilseeds.*

Interview with agricultural sustainability and water management expert, Jan. 2015.

Yet, the politically secured protection of the industry, far from providing a basis for the development of biofuels, has imprisoned the sugarcane economy in its traditional markets for sugar and ethanol for alcohol and industry. The dominant position of sugarcane in Maharashtra has consequently intensified rather than diminished the threats of food insecurity and climate change.

The Future for Biofuels in India

Aside from the entrenched institutional obstacles within agriculture to the development of biofuels, there is the wider context of the gap between legal regulatory frameworks and their implementation in reality, so pervasive in India. There has been little effective institutional or policy support for an emerging biofuels market beyond the proposal of relatively ambitious targets: "*Whatever policy exists, there is no mechanism to make it happen*" (Interview with expert on Indian agriculture and energy policy, Jan. 2015). As with the German *Energiewende*, there is a sense that transport is neglected relative to power generation: "*When we talk about increasing energy efficiency, we mostly understand that everyone is talking about power. Transport is somewhat left out*" (Interview with civil society representative, Jan. 2015). Facing less intractable issues inherent in agricultural, land-holding and use, a structured and systematic effort in solar power has, by contrast, transformed the sector. A National Solar Mission coordinated policymaking and implementation for the solar sector. By contrast, "*the government prepared a Biomass Mission, but it has not seen daylight yet. It is still in the discussion stage and has not been activated... Bioenergy is anchored to many agencies, each supports some niche activities but there is a lack of an overall integrated effort*" (Interview with advance biofuel experts, Jan. 2015). The entire architecture of institutional, financial and marketing support for innovation is thus missing. Consequently, as illustrated in Tables 3.1 and 3.2, in spite of its position as one of the global leaders in the sugar league table, India's biofuel industry has scarcely gotten off the ground.

TABLE 3.1 Bioethanol (Thousands of Barrels per Day)

	2008	2009	2010	2011	2012
China	34.4	37.54931	36.67046	38.85897	43.23598
Germany	10	13	13	13.3	13.37
Brazil	466.2913	449.8178	486.0114	392	402.5
India	5	1.72324	0.86162	6.28981	5.25587
United States	605.566	713.49	867.444	908.6192	875.558

TABLE 3.2 Biodiesel (Thousands of Barrels per Day)

	2008	2009	2010	2011	2012
China	5	10.18432	9.78798	14.68197	15.66421
Germany	55	45	49	57.235	54.7
Brazil	20.05742	27.71055	41.12335	46.05804	46.7
India	0.2	1.29243	1.55091	1.7577	1.98172
United States	44.11307	34	22	63	64

Source: US Energy Information Administration. Downloaded 4.8.2015.

Thus, the yawning gap between political pronouncements, legal targets and mandates, on the one hand and the economic reality on the ground, on the other, has to be placed in a distinctively trilemma Indian context. The huge constraints on India's land and water resources to meet both endemic food poverty and a growing food demand are further magnified by the almost intractable obstacles to agricultural development, including land fragmentation and political corruption. The priority accorded to food security and its current lock-in to economic development based on a fossil-carbon-intensive energy pathway has generated an almost uniquely hostile environment to biofuel development, preventing India from turning sweetness into power (with apologies to Mintz, 1985).

CHINA: SQUEEZED BY LAND SCARCITY, FOOD AND ENERGY SECURITY

Despite some major similarities in terms of environmental resources between India and China, the trilemma manifests itself quite differently in China, shaping its development of renewable energy, including biofuels, and the policy and regulatory frameworks responding to it. There is an even greater scarcity of agricultural land and water, a finitude of national resources for a country that needs to feed 20% of the world's population on less than 10% of its agricultural land (Huang and Rozelle, 2009). China has one-fifth of land to grow food per capita compared with the United States, and less than a quarter compared with Brazil (0.08, 0.5 and 0.37 hectares per capita, respectively; India, 0.13 hectares per capita; World Bank, 2016). With a population anticipated to grow until 2050, and with expanding and changing food demand, especially for meat, the pressures on land use can only intensify (Schneider and Sharma, 2014; Sharma, 2014; Norse et al., 2014). At the same time, China's energy demand is growing substantially, and is the world's top importer of oil since 2014 (EIA, 2015). Within a general perspective of self-reliance, renewable energy resources were therefore at least in part a response to risks to energy security, as enshrined in the 2007 Medium and Long-Term Plan for Renewable Energy (Lewis, 2010; Shiyan et al., 2012a,b). Finally, China has undoubtedly faced problems with severe immediate air and groundwater pollution, which, although indirectly linked, has intensified pressures for climate change mitigation and green development (Angang, 2012). Air quality in the capital concentrates political minds at the center. Groundwater pollution from overuse of nitrogen phosphate fertilizers has the dual effect of immediate soil acidification and water eutrophication (Guo et al., 2010), as well as the longer-term climate change from the powerful GHG, nitrous oxide (Norse and Ju, 2015; Zhang et al., 2013). Consequently, there has been a double dynamic of reducing land productivity and capacity for food production through environmental degradation and increasing immediate and long-term pressures for climate change mitigation.

Given these three distinctively Chinese components of the trilemma tensions, the politics and regulation of biofuels in China have evolved in a distinctive pathway. Compared with the major global players such as the United States and Brazil, China initiated its biofuel program relatively late, with the First Five-Year Plan for Bioethanol and the Special Development Plan for Denatured Fuel Ethanol and Bioethanol Gasoline for Automobiles in 2001. Five substantial biorefineries were constructed between 2001 and 2007 with an ambition to produce 10 million tons of bioethanol. The Renewable Energy Law of 2005 further stimulated the growth of the industry by introducing an E10 blending mandate for nine provinces (Shiyan et al., 2012b). This early period was ambitious in addressing both environmental and energy security aspects of the trilemma, promoting unconstrained growth, and aiming high with production targets of 6 million tons in 2010 rising to 15 million in 2020. However, the contrast with current realities could not be greater. Although now globally ranked third in bioethanol production, China is not in the same league as the top two producers, with less than a 10th of the volume (see Table 3.1). The decisive turning point was the rise in world food prices beginning in 2006, and intensifying during 2007 (Yang et al., 2008). The overriding policy of food security effectively changed the course of China's biofuel development strategy, leading to significant changes in the regulatory framework. Up until 2005, there had been a significant grain surplus in China, the feedstock for biorefineries consisting of stale maize and wheat grain. By the time of the food price spikes, the feedstock had already shifted to fresh annual sources.

In 2007, the Medium and Long-term Development Plan for Renewable Energy reset the parameters for biofuel development in China, both by significantly reducing targets, and, in terms of the trilemma more significantly, eliminating biofuel competition with food from China's limited and stressed land and water resources. The 2007 Plan stated that "biofuel must not compete with grain over land, it must not compete with food that consumers demand, it must not compete with feed for livestock, and it must not inflict harm on the environment" (as quoted in Qiu et al., 2012). The policy was further reinforced by the 2008 Land Administration Law, defining marginal land and land of low agricultural productivity as the only remaining potential resource for biofuel development, and even that remained in doubt if irrigation and fertilizers, used for biofuel feedstocks, could also yield economically viable food crops. Targets were reduced by a third, to 4 million tons by 2010 and 10 million by 2020. However, the constraints on land use and nonfood feedstocks render these targets largely unobtainable, with production of bioethanol still falling below 2 million tons in 2014 (GAIN, 2014). The production of bioethanol from grain feedstocks declined from 150,000 tons in 2008 to 135,000 tons in 2010. It is noteworthy that of all the 2007 renewable energy targets, for solar, wind, biogas, or biomass for power generation, only biofuels is failing to achieve them, or even remotely approach them (Shiyan et al., 2012b).

Nonetheless, whether to address the immediate issues of air pollution, long-term energy security or climate change targets, China is still promoting biofuels. The decline in food grain feedstocks has been partly offset by the developing use particularly of sweet sorghum and cassava (GAIN, 2014). However, even using advanced biorefineries and solid-state fermentation, sweet sorghum presents considerable obstacles as an economically viable feedstock with the potential for major expansion. The experience with cassava is equally problematic, notably in the premier cassava biorefinery plant in Guangxi province, which remains highly subsidized (expert interviews, May 2014). However, in terms of trilemma pressures, cassava presents a clear shift, as China is heavily and increasingly dependent on imported cassava, notably from Vietnam and Thailand for the Guangxi biorefinery (Shiyan et al., 2012b; and interviews, May 2014). If strategic decisions to source biofuel feedstocks from abroad were to match the scale of the shift to imported soy for animal feed from Brazil, some of the land constraints facing China would be circumvented—albeit at the cost of imposing new pressures on land outside the country. Finally, although on an experimental basis, China imported 10,500 tons of bioethanol from the United States, opening the possibility of a major shift in sourcing of biofuels, were the current highly protective tariffs for its own domestic industry reduced, as they already have been since 2012 for ASEAN countries (GAIN, 2014).

China thus presents a distinctive and evolving trajectory of regulation and policy towards biofuels in general, and bioethanol in particular, responding to the trilemma challenges specific to its particular national context: immediate sociotopical impacts of groundwater and air pollution; intensifying pressures on scarce agricultural land and water resources; increasing dependency on imported liquid fossil fuels. There are hints of major geopolitical shifts, similar to the shift to imported soy from Brazil, which could change the terms of the Chinese trilemma. If not, in marked contrast to the huge progress in other forms of renewable energy, including bioenergy, biofuel consumption in China appears doomed to substantially miss its current targets, largely as a consequence of some systemic constraints to domestic production from China's land and water resources. Within those constraints, politically food has taken absolute priority over biofuels, since 2007.

BRAZIL: A TRILEMMA TRAJECTORY DERAILED AT HOME AND BLOCKED FROM ABROAD?

In trilemma terms, Brazil's development of biofuels has evolved in a quite distinctive way and in a unique environmental resource context that broadly can be divided into four phases: the energy security and national development phase (1964–1985); Washington consensus liberalization rebalancing fossil- and biofuels, 1985–2003; biofuel revival and reorientation with Flex-Fuel Vehicles and environmentalism, 2003–2007; global crisis, global market constraints, and retreat from environmentalism (2007–present) (Harvey, 2014). As already suggested, apart from the United States, Brazil stands in a different league to any other country, whether in terms of production or consumption of bioethanol. Ranked second as a producer, in terms of per capita production it is close to US levels, but, more pertinently, in terms of per vehicle production, it leads the world with more than double the quantity of biofuel per vehicle than the United States (80 million registered vehicles, at 249 per 1000 population in Brazil in 2012; 253 million registered vehicles, at 809 per 1000 population in the United States in 2012).

The phasing of bioethanol's developmental trajectory in Brazil in trilemma terms can be presented here only briefly and schematically (for some alternative periodizations see Andrade and Miccolis, 2011; Goldemberg et al., 2004; Lehtonen, 2007; Rosillo-Calle and Cortez, 1998; Zuurbier and van de Vooren, 2008). Dating back to 1931 under President Vargas, Brazil had a long "developmental state" promotion of the sugarcane and bioethanol industries, including promotion of fuel blending, symbolized by the creation of the Instituto do Açúcar e do Álcool (IAA) in 1933 (Johnson, 1983; Barzelay, 1986; Evans, 1979, 1982). Under the military dictatorship (1965–84), this logic of economic development combined with a trilemma imperative of energy security, developing domestic energy resources and reducing dependency on oil imports (De Almeida et al., 2007). The oil-shocks of 1975 and 1979, affecting developing economies most severely, prompted the major policy and regulatory instruments of ProAlcool I and ProAlcool II, successively imposing mandatory blending to 20% bioethanol, guaranteed prices, and the introduction of the 100% hydrous ethanol vehicles. Following the tri-pé (tripod) policy combining the state, national and international capital (Evans, 1979), Petrobras, the major oil producer and distributor, became a central instrument for purchasing and distributing bioethanol. The 100% ethanol-engine vehicles were produced by major global car manufacturers as a consequence of negotiations with the military dictatorship (Puppim de Oliveira, 2002; Rosillo-Calle and Cortez, 1998; Goldemberg, 2008).

Following the fall of the Generals, and especially under President Cardoso (1995–2003), there was first a relative stabilization, followed by limited growth of bioethanol production, still directed almost entirely to the domestic market. Under the Washington consensus, subsidies were reduced, price competition between petrol and biofuels liberalized, and the 100% ethanol car was abandoned. However, the mandated blending of the politically instituted bioethanol market remained, and, with rising oil prices, the policies of energy security and promotion of self-sufficiency continued. So, for almost four decades biofuel policy and regulation in Brazil were developed in response to this dominant imperative, without consideration of competition with food, climate change mitigation, or environmental protection.

Although with continuing high oil prices and import-dependency, the energy security pole of the trilemma exercised continuing pull, under President Lula (2003–11) there was a significant shift in policy and regulatory orientation. The shift was generated by both domestically internal and international dynamics, which centered on environmental protection and climate change mitigation, primarily centered around food, with secondary or consequential impacts on sugarcane cultivation for ethanol. From now on, biofuel production and expansion of land under sugarcane cultivation was framed by these broader environmental policies, especially the Zero Deforestation Policy, and the Better Sugar Initiative (shadowing the Roundtable for Responsible Soy, and the Soya Moratorium) (Wilkinson, 2011). The National Agro-energy Plan (2005) and the National Climate Change Plan (2009) both promoted biofuel expansion in Brazil on the same environmentalist regulatory conditions (La Rovere et al., 2011). In 2007, the government promoted a policy of agroecological zoning, restricting sugarcane expansion in the Amazon and Cerrado, a measure further strengthened by a 2008 São Paulo State law and a 2009 Presidential Decree (Andrade and Miccolis, 2011). Although sugarcane for ethanol continues to occupy a tiny fraction of total arable land, around 1%, expansion into the Mato Grasso *cerrado* and the Pantanal has come under environmentalist focus (Wilkinson and Herrera, 2011). Within this context, bioethanol however has been strongly defended for reducing GHG emissions, as the most efficient climate-change mitigation crop (Brehmer and Sanders, 2009). Above all, in a more democratic vein than his military predecessors, Lula negotiated with major car manufacturers to develop the Flex-Fuel Vehicle, giving consumers and the market the choice over which fuel to fill their tanks. FFVs now account for 90% of the new car market. As a consequence of these various regulatory measures, bioethanol witnessed a further period of rapid expansion.

However, the policy and regulation context for Brazilian biofuels—again in parallel with soya beans for food and animal feed—was increasingly affected by international regulatory measures. By 2006, Brazil had become the largest exporter of biofuels—especially to the United States, where sugarcane bioethanol was already classified as an "advanced biofuel," and Europe. In Europe, the 2009 Renewable Energy Directive and Fuel Quality Directive set standards (tailored to European rapeseed biodiesel) for 35% GHG emissions reduction (Harvey and Pilgrim, 2013). While Brazilian sugarcane ethanol met these standards with ease, there has been a developing framework for sustainability certification, also required for imports from Brazil (Pacini and Strapasson, 2012; Pacini et al., 2013; Wilkinson and Herrera, 2011). More significant than the regulatory measures as such, ever since the Searchinger and Fargione controversies concerning indirect land-use change emerged, the whole market environment for biofuels in Europe has witnessed a significant stagnation and even decline (Pilgrim and Harvey, 2010). Indeed, as discussed earlier in relation to Germany, the RED, by no longer specifying requirements for transport energy, has facilitated a concentration on power generation for climate change mitigation, almost exempting transport energy from further regulatory pressures. The recent sharp and prolonged decline in the price of oil, and the promise, however short term, of fracking and nonconventional oil reserves, has further lifted the pressure on policy to develop a long-term strategy for a renewable alternative to the finite resources of fossil fuel. Therefore, both the European regulatory regime, and the political environment that shaped it, are now impacting negatively on the development of Brazilian bioethanol exports.

This shift in the external environment has more recently been accompanied by a further change of course in the politics of biofuels and transport energy in Brazil, heralding the current phase of crisis and reversal for its bioethanol industry. Since the succession of Dilma Rousseff to the Presidency, the environmentalist biofuel objectives of Lula have largely fallen into abeyance. In its place, a pro-poverty policy of capping the price of petrol, and at the same time promoting easy finance for car purchases, has depressed investment in the industry. In the context of Flex-Fuel Vehicles, this pro-petrol policy has intensified competition between fossil- and biofuels, undermining confidence in the future for bioethanol. Unquestionably, the earlier discovery of the major deep ocean Tupi oilfield reserves had already diminished gravitational pull of the fossil-energy security pole of Brazil's trilemma challenges, opening the possibility for greater reliance on petrol. Therefore, although exacerbated by the financial crisis, the credit crunch, as well as a certain complacency in the biorefinery and sugarcane agricultural sectors, the current political environment could scarcely diverge more strongly from the earlier vision of a future green bioeconomy for transport energy. The continuing commercial and political scandals surrounding Petrobras, the single most significant quasimarket actor with a split personality, having a foot in both petrol and biofuel camps, only darkened the horizon further. There are certainly some small signs of innovation, especially the development of coproduction of ethanol from lignocellulosic sugarcane biomass, partly supported by the US mandated market for 2G biofuel, but these are more against than with the present political grain.[1]

Reviewing the long trajectory of Brazilian bioethanol regulation and policy in a trilemma analytical perspective, a number of features stand out. Firstly, given Brazil's vast land and water resources and globally important biodiversity heritage, environmentalist regulation has focused quite distinctively on preventing the extensification of land under cultivation. By contrast, beef production on existing pastures and intensification of cattle rearing, along with beef consumption, amongst the highest in the world, is growing without raising climate change policy or other environmentalist concerns to the same degree. Secondly, and relatedly, food–fuel competition, either in terms of food prices or in terms of competition for land and water, has not emerged as a consideration in the formation of policy and regulation during the whole Brazilian evolution of biofuels, whether bioethanol or biodiesel, including from soy. Finally, although Brazil is better placed and more advanced than any other country to develop a strategy for the long-term substitution of renewable biofuels for fossil transport fuel (Mathews, 2007), the intermingling and near fusion of the political apparatus with Petrobras, a global oil major, no doubt obstructs the innovation pathway towards that end. The long history of Petrobras' involvement with biofuels and oil has been a testimony to its strategic ambivalence (Barzelay, 1986), now reinforced by the discovery of new oilfield resources which has shifted the gravitational pull of the different poles of Brazil's trilemma away from biofuels.

FINAL REMARKS

In the heyday of biofuel optimism, there was a vision of a new global South–North geopolitical pact for terrestrial transport, with a prospect of "18 Brazils" across the subtropical world providing a substantial energy contribution, both renewable and environmentally beneficial (Mathews, 2007). Less than a decade later, that vision has dimmed, but without any significant alternative to the dominance of conventional fuels in transport. In the interval, two major energy resources—land and fossil fuels—have witnessed a political recasting of their natural finitudes. Now, rather than being viewed in terms of what is naturally available, both have become increasingly subject to political limits for their exploitation. There are scientific arguments that call for 80% of existing fossil carbon to remain unexploited in order to keep global warming below 2°C (Leaton et al., 2013). Deep ocean and polar region exploitation are deeply contentious and fracking is prohibited in some countries, such as in Germany. Peak oil is becoming, if to a limited extent, "politically capped" peak oil. In

---

1. Notably, the Granbio biorefinery in Alagoas State in northern Brazil, and the Riazen plant operated jointly by Shell and Cosan. Interview with Carlos Labate, ESALQ, Nov. 20, 2014.

similar fashion, and above all in Brazil, legal and self-regulatory measures are restricting use of land for agriculture, and the controversies surrounding land-use change are similarly developing constraints to the exploitation of a natural resource (energy crops) across the globe.

The purpose of this chapter has been to show how the different poles of the trilemma (food, energy and climate change) particularly as mediated by land, have a different and shifting gravitational pull in different countries. Partly as a consequence of their own environmental resources, and partly because of their political systems and political orientations, the politics and regulation of biofuels have varied significantly as a consequence of the strength of gravitational pull.

Brazil stands out as uniquely endowed with land and sun environmental resources, but with its world-leading intensity of bioethanol production and consumption now faced with a new context of potential oil resources from deep ocean reserves. In contrast, India and especially China, are confronted with global extremes of agricultural land and water scarcity. Germany, long dependent on importing fossil carbon energy, and with an advanced industrial agriculture, is now both dynamically shifting towards renewable energy power generation, and environmentally concerned about its indirect claims on pristine land outside its own territorial boundaries. Both India and China are dominated by the politics of food security, but whereas India struggles to pursue a trajectory of food self-sufficiency, China has increasingly shifted towards claims on extraterritorial land, above all in Brazil, for its food supply chains.

Given these contrasting underlying gravitational weights to the three poles of the trilemma, the impacts of two global shifts have played out in very different ways in our four cases, firstly the dramatic drop in oil prices following the global economic slowdown and the unconventional oil and gas boom in the United States; and secondly, the food price spikes of 2007–08. It is clear that the transport energy security pressure has been relaxed in the short term, especially for the most oil-import-dependent economies. In Brazil, within the context of low global oil prices, a pro-poverty pro-oil politics of recent years has contributed to the negative environment for further biofuel innovation and development. By contrast, the food price spikes have had lasting effects on biofuel policies in China and India.

As a consequence of these trilemma dynamics, there have been quite dramatic shifts in regulation and biofuel developments, in contrasting ways, in both Germany and China, rapid growth followed by reversals. In India, political paralysis, entrenched interests and institutional barriers have left biofuels standing at the starting gate, with state incapacity to respond to the challenges faced, within the environmental constraints of land and water. Brazil, in its uniquely favored position, without dramatic policy shifts appears to have lost its long-term strategic advantage to becoming the world's greenest energy economy.

Of course, it is not being claimed that the dynamics of the trilemma's shifting gravitational poles and their contrasting effects on policy and regulation for biofuels is a total explanation of the different strategies being pursued in our four cases. The contrasting political systems and state capacities undoubt-edly play a greater role than discussed here. Nonetheless, the interlocking tri-lemma challenges of food, energy and transport energy, however fluctuating, will undoubtedly intensify rather than diminish in the long term, with contrast-ing consequences and dynamics in the interactions between political economies and their resource environments in their search for renewable and sustainable energy resources for transport. On the one hand, the political and regulatory shifts constricting biofuel development have so far increased oil dependency in the absence of any comprehensive alternative from technologies such as electric or hydrogen cells, to meet the multiple needs of terrestrial mobility for both heavy goods and light vehicles. On the other, the global vision of "18 Brazils" powering Flex-Fuel Vehicles across the world was predicated on an illusion of a single technological fix similar to the petrochemical complex of the 20th century, ignoring the different trilemma challenges and political contexts faced by different countries with their starkly varied access to planetary resources. Neither India, nor China could become an nth Brazil—and Germany barred the route to importing biofuels from the global South. Meanwhile, trilemma chal-lenges remain and grow.

ACKNOWLEDGMENT

The research for this chapter was funded by an ESRC Professorial Fellowship held by Prof Mark Harvey, Ref. ES/K010530/1.

REFERENCES

Ahn, S., Graczyk, D., 2012. Understanding Energy Challenges in India: Policies, Players and Issues. IEA, Paris.

Andrade, R.M.T., Miccolis, A., 2011. Policies and Institutional and Legal Frameworks in the Expansion of Brazilian Biofuels. Working Paper 71. CIFOR.

Angang, H., 2012. China: Innovative Green Development. Institute for Contemporary China Studies. Tsinghua University.

Barzelay, M., 1986. The Politicized Market Economy. Alcohol in Brazil's Energy Strategy. University of California Press, Berkeley, CA.

Bharucha, Z.P., 2014. The Brazil-China Soy Complex: A Global Link in the Food-Energy-Climate Change Trilemma. CRESI Working Paper 2014-01. University of Essex.

Block, F., 2007. Understanding the diverging trajectories of Europe and the United States: a neo-Polanyian analysis. Politics Society 35 (1), 1–31.

Block, F., 2008. Swimming against the current: the rise of a hidden developmental state in the United States. Politics Society 36 (2), 169–2006.

Borras, S.M., McMichael, P., Scoones, I., 2011. The Politics of Biofuels, Land and Agrarian Change. Routledge, London.

Brehmer, B., Sanders, J., 2009. Assessing the current Brazilian sugarcane industry and directing developments for maximum fossil fuel mitigation for the international petrochemical market. Biofuels Bioprod. Bioref. 3, 347–360.

Datta, K.K., de Jong, C., 2002. Adverse effect of waterlogging and soil salinity on crop and land productivity in northwest region of Haryana, India. Agric. Water Manage. 57 (3), 223–238.

De Almeida, E.F., Bomtempo, J.V., De Souza e Silva, C.M., 2007 The Performance of Brazilian Biofuels: An Economic, Environmental and Social Analysis. Federal University of Rio de Janeiro, Working Paper, 5.

De Sousa, P.T., Dall'Oglio, E.L., 2008. The ethanol and biodiesel programmes in Brazil. In: Mytelka, L.K., Barnes, G. (Eds.), Making Choices About Hydrogen: Transport Issues for Developing Countries UNU Press/IDRC, Canada, pp. 118–140.

Doornbosch, R., Steenblick, R., September 11–12, 2007. Biofuels: Is the Cure Worse Than the Disease? OECD Round Table on Sustainable Development, Paris.

D'Souza, R., 2008. Framing India's hydraulic crisis: the politics of the modern large dam. Monthly Rev. 60 (3) <http://monthlyreview.org/2008/07/01/framing-indias-hydraulic-crisis-the-politics-of-the-modern-large-dam/>.

EIA, 2015. China. URL: https://www.eia.gov/beta/international/analysis.cfm?iso=CHN (accessed 24.03.16).

Eickhout, B., February 20, 2015. EU must tackle indirect land-use change caused by biofuels. Parliament Mag. <https://www.theparliamentmagazine.eu/articles/opinion/eu-must-tackle-indirect-land-use-change-caused-biofuels>.

European Commission, 2011. Roadmap to a Single European Transport Area—Towards a Competitive and Resource Efficient Transport System. European Commission, Brussels.

European Environment Agency, 2015. Greenhouse gas emissions from transport. URL: <http://www.eea.europa.eu/data-and-maps/indicators/transport-emissions-of-greenhouse-gases/transport-emissions-of-greenhouse-gases-5/> (accessed 24.03.16).

European Commission, 2015a. Reducing Emissions From Transport. Available from: <http://ec.europa.eu/clima/policies/transport/index_en.htm>.

European Commission, June 15, 2015b. Technical Assessment of the EU Biofuel Sustainability and Feasibility of 10% Renewable Energy Target in Transport. European Commission, Brussels, <https://ec.europa.eu/energy/sites/ener/files/documents/SWD_2015_117_F1_OTHER_STAFF_WORKING_PAPER_EN_V5_P1_814939.PDF>.

Evans, P., 1979. Dependent Development: The Alliance of Multinational, National and Local Capital in Brazil. Princeton University Press, Princeton, NJ.

Evans, P., 1982. Reinventing the bourgeoisie: state entrepreneurship in dependent capitalist development. Am. J. Sociol. 88 Suppl., 210–247.

FAO, 2015. The State of Food Insecurity in the World. FAO, Rome.

FAOSTAT, 2015. Production/Crops, Sugarcane. (accessed 24.03.16).

Fischer, G., Teixeira, E., Hizsnyik, E.T., Velthuizen, H.V., 2008. Land use dynamics and sugarcane production (Chapter 2) Sugarcane Ethanol: Contribution to Climate Change Mitigation and the Environment. Wageningen Academic, Wageningen.29–62

GAIN (Global Agriculture Information Network), 2014. Biofuels Annual Report. People's Republic of China. CH14038. USDA.

Gee, S., McMeekin, A., 2011. Eco-innovation systems and problem sequences: the contrasting cases of US and Brazilian biofuels. Ind. Innov. 18 (3), 301–316.

Gibbs, H.S., Rausch, L., Munger, J., Schelly, I., Morton, D.C., Noojipaddy, P., et al., 2015. Brazil's Soy Moratorium. Science 347 (6220), 377–378.

Goldemberg, J., 2008. The Brazilian biofuels industry. Biotechnol. Biofuels 1 (6), 1–7.

Goldemberg, J., Coehlo, S.T., Nastari, P.M., Lucon, O., 2004. Ethanol learning curve—the Brazilian experience. Biomass Bioenergy 26, 301–304.

Guo, J.H., Liu, X.J., Zhang, Y., Shen, J.L., Han, W.X., Zhang, W.F., et al., 2010. Significant acidification in major Chinese croplands. Science 327 (5968), 1008–1010.

Harvey, M., 2014. On the horns of the food-energy-climate change trilemma: towards a socio-economic analysis, *Theory, Society and Culture*. Energy Society 31 (5), 155–182. (special issue).

Harvey, M., McMeekin, S., 2010. The Political Shaping of Transitions to Biofuels in Europe, Brazil, and the USA. CRESI Working Paper, 2010-02. <http://repository.essex.ac.uk/2296/1/CWP-2010-02-Political-Shaping-Final.pdf>.

Harvey, M., Pilgrim, S., 2011. The new competition for land: food, energy and climate change. Food Policy J. 36 (S1), 40–51.

Harvey, M., Pilgrim, S., 2013. Rudderless in a sea of yellow: the European political economy impasse for renewable transport energy. New Pol. Econ. 18 (3), 364–390.

Hill, N., Brannigan, C., Smokers, R., Schroten, A., van Essen, H., Skinner, I., 2012. EU Transport GHG: Routes to 2050 II. Developing a better understanding of the secondary impacts and key sensitivities for the decarbonisation of the EU's transport sector by 2050. Final project report produced as part of a contract between European Commission Directorate-General Climate Action and AEA Technology plc. <http://www.eutransportghg2050.eu/cms/assets/Uploads/Reports/EU-Transport-GHG-2050-II-Final-Report-29Jul12.pdf>.

Huang, J., Rozelle, S., 2009. Agricultural Development and Nutrition: The Policies Behind China's Success. World Food Programme.

ICCT, 2015. India. Available from: <http://www.theicct.org/india>.

ICIS, December 15, 2014. Indian oil firms stops ethanol procurement from sugar mills. ICIS News Available from: <http://www.icis.com/resources/news/2014/12/15/9846528/indian-oil-firms-stops-ethanol-procurement-from-sugar-mills/#>.

IEA (International Energy Authority), 2012. World Energy Outlook 2012. IEA, Paris.

IEA, 2012. Oil Market Report. OECD/IEA, Paris.

IEA, 2014. World Energy Outlook 2014. IEA, Paris.

Jha, D.K., 2014. OMCs cancel ethanol procurement tender. Business Standard December 2, 2015. Available from: <http://www.business-standard.com/article/companies/omcs-cancel-ethanol-procurement-tender-114120200881_1.html>.

Johnson, F.I., 1983. Sugar in Brazil: policy and production. J. Dev. Areas 17 (2), 243–256.

Krebs, J.R., Wilson, J.D., Bradbury, R.B., Siriwardena, G.M., 1999. The second silent spring? Nature 400 (6745), 611–612.

Langniss, O., Diekmann, J., Lehr, U., 2009. Advanced mechanisms for the promotion of renewable energy—models for the future evolution of the German Renewable Energy Act. Energy Policy 37 (4), 1289–1297.

La Rovere, E.L., Pereira, A.S., Simões, A.F., 2011. Biofuels and sustainable energy development in Brazil. World Dev. 39 (6), 1026–1036.

Leaton, J., Ranger, N., Ward, B., Sussams, L., Brown, M., 2013. Unburnable Carbon 2013: Wasted capital and stranded assets Carbon Tracker and Grantham Research Institute on Climate Change and the Environment. London School of Economics, London, <http://www.carbontracker.org/wastedcapital>.

Lehtonen, M., 2007. Biofuel Transitions and Global Governance: Lessons From Brazil. Conference Paper, Amsterdam. <http://www.2007amsterdamconference.org/Downloads/AC2007_Lehtonen.pdf>.

Lewis, J.I., 2010. The evolving role of carbon finance in promoting renewable energy development in China. Energy Policy 38 (6), 2875–2886.

Lieven, T., Mühlmeier, S., Henkel, S., Waller, J.F., 2011. Who will buy electric cars? An empirical study in Germany. Transp. Res. Part D Transp. Environ. 16 (3), 236–243.

Luthra, S., 2014. India's Shift to a Sustainable Energy Future. World Resources Institute. <http://www.wri.org/blog/2014/03/indias-shift-sustainable-energy-future>.

Mathews, J.A., 2007. Biofuels: what a biopact between North and South could achieve. Energy Policy 35 (7), 3550–3570.

Mintz, S.W., 1985. Sweetness and Power. The Place of Sugar in Modern History. Penguin, London.

Norse, D., Lu, Y., Huang, J., 2014. China's food security: is it a national, regional or global issue? In: Brown, K. (Ed.), China and the EU in Context: Insights for Business and Investors Palgrave, London.

Norse, D., Ju, X., 2015. Environmental costs of China's food security. Agric. Ecosyst. Environ. 209, 5–14.

Pacini, H., Strapasson, A.B., 2012. Innovation subject to sustainability: the European policy on biofuels and its effects on innovation in the Brazilian bioethanol industry. J. Contemp. Eur. Res. 8 (3), 367–397.

Pacini, H., Assunção, L., Van Dam, J., Toneto, R., 2013. The price for biofuels sustainability. Energy Policy 59, 898–903.

Palmer, J.R., 2014. Biofuels and the politics of land-use change: tracing the interactions of discourse and place in European policy making. Environ. Plann. A 46, 337–352.

Pilgrim, S., Harvey, M., 2010. Battles over biofuels in Europe: NGOs and the politics of markets. Sociol. Res. Online 15 (3), 4.

Piorr, A., Ungaro, F., Ciancaglini, A., Happe, K., Sahrbacher, A., Sattler, C., et al., 2009. Integrated assessment of future CAP policies: land use changes, spatial patterns and targeting. Environ. Sci. Policy 12 (8), 1122–1136.

Planning Commission, Government of India, 2002. India Vision 2002. Available from: <http://planningcommission.nic.in/reports/genrep/pl_vsn2020.pdf>.

Planning Commission, Government of India, 2014. Power and Energy. Available from: <http://planningcommission.nic.in/sectors/index.php?sectors=energy>.

Polanyi, K., 2015. General economy history. In: Resta, G., Catanzariti, M. (Eds.), For a New West Polity, London.

Pradhan, S., Ruysenaar, S., 2014. Burning desires: untangling and interpreting 'pro-poor biofuel policy processes in India and South Africa'. Environ. Plann. A 46, 299–317.

Pretty, J., Bharucha, Z.P., 2014. Sustainable intensification in agricultural systems. Ann. Bot. 114 (8), 1571–1596.

Puppim de Oliveira, J.A., 2002. The policymaking process for creating competitive assets for the use of biomass energy: the Brazilian alcohol programme. Renewable Sustainable Energy Rev. 6, 129–140.

Qiu, H., Sun, L., Huang, J., Rozelle, S., 2012. Liquid biofuels in China: current status, government policies, and future opportunities and challenges. Renewable Sustainable Energy Rev 16, 3095–3104.

Rauch, A., Thöne, M., 2012. Biofuels—At What Cost. Mandating Ethanol and Biodiesel Consumption in Germany. GSI, Institute of Sustainable Development, Köln.

Ray, S., Goldar, A., Miglani, S., December 2011. Ethanol Blending Policy: Issues Related to Pricing. Indian Council for Research on International Economic Relations Policy Series No 10, New Delhi, India.

Rodell, M., Velicogna, I., Famiglietti, J.S., 2009. Satellite-based estimates of groundwater depletion in India. Nature 460, 999–1002.

Rosillo-Calle, F., Cortez, L.A.B., 1998. Towards ProAlcool II—a review of the Brazilian Bioethanol Programme. Biomass Bioenergy 14 (2), 115–124.

Roy, S., August 13, 2015. UAE trip: Infra funds, oil trade on PM Modi's agenda. Indian Express Available from: <http://indianexpress.com/article/india/india-others/uae-trip-infra-funds-oil-trade-on-pm-modis-agenda/>.

Ryan, J.G., Spencer, D.C., 2001. Future Challenges and Opportunities for Agricultural R&D in the Semi-Arid Tropics. International Crops Research Institute for the Semi-Arid Tropics (ICRISAT), Patancheru, India.

Schneider, M., Sharma, S., 2014. China's Pork Miracle? Agribusiness and Development in China's Pork Industry. Institute for Agriculture and Trade Policy (IATP), Washington, DC.

Sharma, S., 2014. The Need for Feed. China's Demand for Industrialised Meat and Its Impacts. Institute for Agriculture and Trade Policy (IATP), Washington, DC.

Shiyan, C., Lili, Z., Timilsina, G.R., Xiliang, Z., 2012a. Biofuels development in China: technology options and policies needed to meet the 2020 target. Energy Policy 51, 64–79.

Shiyan, C., Lili, Z., Timilsina, G.R., Xiliang, Z., 2012b. Development of Biofuels in China. Technologies, Economics and Policies. Policy Research Working Paper 6243, World Bank, Development Research Group, Environment and Energy Team. pp. 1–34.

Smith, P., et al., 2010. Competition for land. Phil. Trans. R. Soc. 365, 2941–2957.

Stattman, S.L., Hospes, O., Mol, A.P., 2013. Governing biofuels in Brazil: a comparison of ethanol and biodiesel policies. Energy Policy 61, 22–30.

Strunz, S., 2014. German energy transition as an example of a regime shift. Energy Policy 100, 150–158.

Taylor, J.B., 2009. Getting Off Track. How Government Actions Caused, Prolonged and Worsened the Financial Crisis. Hoover Institution Press, Stanford, CA.

Tiwari, V.M., Wahr, J., Swenson, S., 2009. Dwindling groundwater resources in northern India, from satellite gravity observations. Geophys. Res. Lett. 36, L18401. http://dx.doi.org/10.1029/2009GL039401

Turner, A., 2013. The Indian Monsoon and Climate Change. Walker Institute for Climate System Research Walker Institute, University of Reading. <http://www.walker-institute.ac.uk/publications/factsheets/new2013/Walker%20Institute%20Indian%20monsoon.pdf>.

UFOP (Union for the Advancement of Oil and Protein Plants), 2013a. Biodiesel 2012-2013. Report on the Current Situation and Prospects. <www.ufop.de>.

UFOP (Union for the Advancement of Oil and Protein Plants), 2013b. Rapeseed—Opportunity or Risk for the Future. <www.ufop.de>.

Urban Emissions.info, 2012. Road Transport in India 2010-2030—Emissions, Pollution and Health Impacts. Available from: <http://www.urbanemissions.info/india-road-transport>.

USDA, 2014. India: Biofuels Annual 2014. USDA GAIN Report No. IN4055. 7/1/2014. Available from: <http://gain.fas.usda.gov/Recent%20GAIN%20Publications/Biofuels%20Annual_New%20Delhi_India_7-1-2014.pdf>.

US Energy Information Administration, 2015. International Energy Statistics: Production of Crude Oil including Lease Condensate 2014. US Department of Energy. www.eia.gov. (accessed 24.03.16).

Wilkinson, J., 2011. From fair trade to responsible soy: social movements and the qualification of agrofood markets. Environ. Plann. A 43, 2012–2016.

Wilkinson, J., Herrera, S., 2011. Biofuels in Brazil: debates and impacts. In: Borras, S.M., McMichael, P., Scoones, I. (Eds.), The Politics of Biofuels, Land and Agrarian Change Routledge, London, pp. 175–194.

World Bank, 2016. Arable land (hectares per person). URL: <http://data.worldbank.org/indicator/AG.LND.ARBL.HA.PC/> (accessed 24.03.16).

Yang, J., Qui, H., Huang, J., Rozelle, S., 2008. Fighting global food prices in the developing world: the response of China and its effect on domestic and world markets. Agric. Econ. 39, 453–464.

Zhang, F., Chen, X., Vitousek, P., 2013. Chinese agriculture: an experiment for the world. Nature 497 (7447), 33–35.

Zuurbier, P., van de Vooren, J. (Eds.), 2008. Sugarcane Ethanol. Contributions to Climate Change Mitigation and the Environment Wageningen Academic Publishers, Wageningen.

Chapter 4

Innovation Systems of Ethanol in Brazil and the United States: Making a New Fuel Competitive

L.C. de Sousa[1], N.S. Vonortas[2,3,4], I.T. Santos[5] and D.F. de Toledo Filho[1]
[1]*Ministry of Development, Industry and Foreign Trade, Brazil* [2]*CISTP and Department of Economics, The George Washington University, Washington, DC, United States*
[3]*São Paulo Excellence Chair (SPEC) in technology and innovation policy, University of Campinas (UNICAMP), São Paulo, Brazil* [4]*ISSEK, Higher School of Economics (HSE), National Research University, Moscow, Russia* [5]*Center for Sustainability Studies of São Paulo School of Business Administration (GVces/EAESP), Getulio Vargas Foundation (FGV), São Paulo, Brazil*

INTRODUCTION

Modern forms of energy are one of the pillars of society today and are also largely responsible for the huge economic improvement that has occurred since the Industrial Revolution. Major international organizations consider that energy poverty is one of the main obstacles to sustained economic development. These reasons, coupled with the complex technical issues involved, make the energy sector one of the most regulated economic sectors in the world. Despite the advantages and importance of tighter regulation, one of the problems this creates is that innovation in the energy segment becomes more difficult than in other economic sectors. More radical changes, such as adding a new fuel for transportation, come to depend not only on market conditions but also on political decisions and regulatory bodies.

In addition to regulation, a serious challenge for biofuels is that the average consumer does not place additional value to the availability of different types of fuels. Similarly for environmental concerns: the average consumer does not yet accept to pay more for biofuel that mitigates the emission of pollutants. In the short and medium term, with low prices and very significant quantities of proven reserves of fossil fuels, there is a huge challenge to increase the use of biofuels just through market incentives given their generally higher prices compared to competing fossil fuels.

Global Bioethanol.

Another complication associated with the increased use of biofuels is the large role of the energy sector in the economy, influencing virtually every other sector, thus being important across a variety of policies, such as energy, agricultural, environmental and industrial policies. Brazil initiated its ProAlcool policy in the 1970s as a response to the first major oil shock and the associated difficulties with the country's external trade balance. This external economic shock added a strong argument to the existing industrial policy orientation of the time (import substitution) strengthening the case in favor of homegrown ethanol. More recently, the main motivations of the United States to begin its own biofuel program were the environment and security in terms of lessening dependence on perceived unstable political regimes in major oil- and gas-producing countries in the Middle East and elsewhere. While the two countries had partially different reasons for supporting biofuels, they shared a very strong economic security incentive that greatly played along the interests of another major traditional sector: agriculture. Even with different initial drivers, agriculture has been heavily impacted by the biofuel programs in both countries.

Ethanol tends to be more expensive than gasoline except in special circumstances and locations. Besides the price, several other difficulties have hampered the expansion of ethanol use: (1) the difficulty of demonstrating that the environmental benefits justify the public effort; (2) the debate on the impact of ethanol production on food prices; (3) the advent of new alternative transportation technologies, such as electric cars; (4) the technological development of the sector which has tended to be slower than expected; and (5) continuing subsidies to fossil fuels. While coping with all these issues is already extremely difficult, there is yet an additional challenge: policy coordination. The decision *loci* of industrial, technological, energy and agricultural policies are different, as is their timeframe. This makes the use of ethanol in its two largest producers and consumers (Brazil and the United States) a policy of ups and downs, with periods of high growth followed by periods of stagnation.

This chapter aims at juxtaposing the innovation systems for ethanol in Brazil and the United States. This juxtaposition will help evaluate the main strengths and weaknesses of each country, thus allowing to indicate the parts of the system that can be improved in each case. The comparison will be based on the theoretical framework of innovation systems, specifically using the approach of innovation functions, initially developed by Hekkert et al. (2007). The chapter starts with a brief description of innovation functions and then presents specific characteristics of the Brazilian and US systems emphasizing aspects of (1) the coordination of energy policy with industrial and agricultural policies, (2) environmental aspects of biofuels that argue for (or against) their deployment, and (3) regulation of biofuels. Finally, the chapter presents a general comparison of both countries regarding strengths and weaknesses relevant to the development of second-generation ethanol technologies (2G ethanol).

TECHNOLOGY INNOVATION SYSTEM AND SYSTEM FUNCTIONS

Hekkert et al. (2007) consider that the systems approach to evaluating innovation presents two problems for the understanding of technological change. Firstly, while the theory is based on studies of interactive learning and evolutionary economics, most analyses are quasistatic. The focus is on the comparison of the social structure of different innovation systems (actors, their relationships and institutions) in order to explain performance differences. Little emphasis is given to the analysis of innovation system dynamics. Secondly, it is pointed out that the explanatory power of the model is mainly in institutions (macro level) and less on entrepreneurial action (micro level). In order to assess technological change these authors propose the use of a more precisely cut approach, which they call the technology innovation system (TIS). The TIS includes all agents and technologies that are specific to the technological area under investigation. It involves agents from different countries and includes various sectoral systems, depending on the technology used. The purpose is to reduce the number of agents and relations, compared with that of national systems, which are much broader and more complex to map.

In order to delineate the specific innovation system under investigation, Hekkert et al. (2007) identify the main activities related to the system's success and how they can be translated into innovation functions. The analysis of functions is carried out through an extensive mapping of the events that are taking place within the TIS and change its configuration. The system functions (SF) proposed by Hekkert et al. (2007) are listed below. Table 4.1 provides a list of proposed indicators for assessment of each function.

1. *Entrepreneurial activities*: Are entrepreneurs linking knowledge to market opportunities central to innovation? The function involves the ability to take risks and "learning by experimentation." Entrepreneurs may be new entrants or existing businesses that diversify their strategy to take advantage of new developments.
2. *Knowledge development*: This function is directly related to "learning by searching," "learning by doing" and to research and development (R&D) activities. This function draws on Lundvall (1992) who states that "the most fundamental resource in modern economy is knowledge, and, accordingly, that the most important process is learning." The development of knowledge is a prerequisite for the TIS.
3. *Knowledge diffusion*: This function occurs on a network of actors who are constantly exchanging experiences and incorporate "learning by interacting" and "learning by using." Policy decisions should be consistent with the latest technological developments and R&D agendas affected by norms and values. This function aims to identify how knowledge flows within the TIS.

TABLE 4.1 Functions of Innovation Systems (SFs)

Functions	Examples of Indicators	Event Categories (Positive/Negative)
Entrepreneurial activities	<ul><li>Experimental projects</li><li>New product introductions</li><li>New businesses</li></ul>	<ul><li>Projects (initiated/stopped)</li><li>Turn-key technology</li><li>Lack of contractors</li></ul>
Knowledge development	<ul><li>Scientific publications</li><li>Learning by searching</li><li>Learning by doing</li><li>Technology application (learning by using)</li><li>Learning curves</li><li>R&D projects</li><li>Patents</li></ul>	<ul><li>Desktop-, assessment-, feasibility studies</li><li>Reports</li><li>R&D projects</li><li>Patents</li></ul>
Knowledge diffusion	<ul><li>Collaboration patterns</li><li>Demonstration projects</li><li>Knowledge and experience networks</li><li>Conferences and debate meetings</li><li>Interest organizations (industrial, environmental, etc.)</li></ul>	<ul><li>Conferences</li><li>Workshops</li><li>Platforms</li><li>Guidance</li></ul>
Search guidance	<ul><li>Policy action plans</li><li>Shared strategies and roadmaps</li><li>Debate activities</li></ul>	<ul><li>Expectations on renewable energy</li><li>Government regulation on renewable energy</li></ul>
Market formation	<ul><li>Market application, market shares</li><li>Public market support</li><li>Niche markets</li><li>Standards, certifications</li><li>Trade, exports</li><li>Environmental impacts</li></ul>	<ul><li>Feed-in rates, environmental standards, green labels</li></ul>
Resource mobilization	<ul><li>R&D funding</li><li>Investments</li><li>Personnel: R&D and general employment</li></ul>	<ul><li>Subsidies</li><li>Investments</li></ul>
Legitimacy creation	<ul><li>Public opinions on energy technologies and systems</li><li>Regulatory acceptance and integration</li></ul>	<ul><li>Lobby by agents to improve technical, institutional, financial conditions</li><li>Expressed lack of lobby agents</li><li>Lobby for other technology that competes with particular technology</li><li>Resistance to change by neighbors</li></ul>

Source: Borup, M. et al., 2013. Indicators of Energy Innovation Systems and Their Dynamics. A Review of Current Practice and Research in the Field. Disponível em: <www.eis-all.dk>(Borup et al., 2013), based on Hekkert, M.P., Suurs, R.A.A., Negro, S.O., Kuhlmann, S., Smits, R.E.H.M., 2007. Functions of innovation systems: a new approach for analysing technological change. Technol. Forecast. Social Change 74, 413–432 (Hekkert et al., 2007); Bergek, A. et al. Analyzing the functional dynamics of technological innovation systems: A scheme of analysis. Research Policy, v. 37, n. 3, pp. 407–429, abr. 2008. (Bergek et al., 2008); and Hekkert, M.P.; Negro, S.O. Functions of innovation systems as a framework to understand sustainable technological change: Empirical evidence for earlier claims. Technological Forecasting and Social Change, v. 76, n. 4, pp. 584–594, 2009 (Hekkert and Negro, 2009).

4. *Search guidance*: This function relates to the definition of policy goals and of a research agenda that selects the best technology from the available options. It is also related to political decisions such as the share of renewables in the future energy mix. The efficient management of resources is considered essential for the success of the TIS. The function is also related to the fact that technological change is not autonomous but rather it is embedded in the institutional and political environment.

5. *Market formation*: This function relates to the temporary creation of niche markets, allowing new technologies to penetrate the business environment. It can also be achieved through the creation of temporary competitive advantage using taxes and favorable rates or minimum consumption quota regimes.

6. *Resource mobilization*: This function relates to the human and financial resources necessary for the development of a technology or product over time. Resources are both private and public and preferably should be complementary, aiming to increase the synergies between them.

7. *Legitimacy creation*: This function serves to overcome resistance to change. It can be performed by groups advocating the new technology.

SFs influence each other: the fulfillment of a given function probably has effects on the performance of other functions. These multiple interactions are nonlinear and can positively or negatively affect the system. Fig. 4.1 is borrowed from Hekkert et al. (2007) and shows three typical changes in a TIS. A common trigger for virtuous cycles is the direction of the search (C), leading to the increase of resources invested, the creation and diffusion of knowledge, and to increased expectations which could generate new business activity, strengthen legitimacy and enhance markets. These interactions would be reinforced over time, strengthening the system. Another possible start of a virtuous cycle could be entrepreneurs lobbying successfully for better economic conditions (legitimacy) which could result in increased expectations (B).

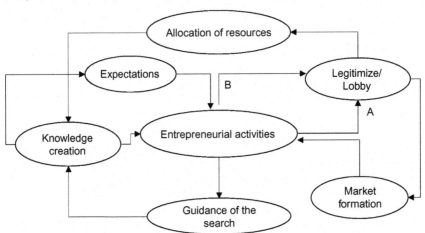

FIGURE 4.1 Technology innovation system (TIS): three typical changes. *Hekkert, M.P., Suurs, R.A.A., Negro, S.O., Kuhlmann, S., Smits, R.E.H.M., 2007. Functions of innovation systems: a new approach for analysing technological change. Technol. Forecast. Social Change 74, 426.*

TABLE 4.2 Symbols Used in the Graphic Representation of Innovation Functions

RFS2	Indicates an action or an identified fact that occurred before the reporting period
High oil price	Indicates an action or a fact that is macroeconomic or external to the industry (defined closely)
Biotechnology developments in other countries	Indicates an action or a fact that occurs in another country and affects the sector also in the country's system under review
Ethanol infrastructure grants and loan guarantees	Indicates a Federal or State Public action
E2G premium price on USA	Indicates the result of an action or relevant fact identified
Market formation / Ethanol 1G / Ethanol 2G / Bioeletricity	Indicates the innovation function involved. The analysis was divided into first-generation (1 G) and second-generation (2 G) ethanol and bioenergy from biomass
Some industrial projects	Indicates the outcome of the TIS under review
→	Indicates a positive relationship
→	Indicates a negative relationship

A TIS is successful when all major components are present. It is important to highlight that the SFs are complementary between different TIS and highly interdependent. The lack of one of the SFs can become a bottleneck for the whole TIS. It is exactly from this point that one can identify the areas that need to be better developed. Having one strong SF is not adequate for the development of the TIS. It is necessary that all SFs work together to produce the desired results.

This chapter puts forward a graphical representation of the ethanol TIS. The symbols used are shown in Table 4.2.

Considering the complexity of the ethanol TIS, we have restricted the mapping to the events considered most relevant. Fig. 4.2 presents an example of graphic representation used. The development of flex-fuel technology, which predates the reporting period, giving conditions to mobilize resources in the form of tax reduction, positively affects market formation. Increased labor costs and land prices accelerate the process of mechanization, which results in decreased sugarcane productivity per hectare and negatively affects the market formation as it increases the cost of ethanol. The result of these two factors is market stagnation.

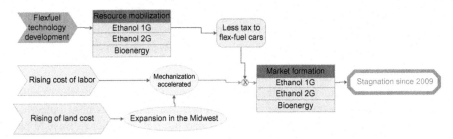

FIGURE 4.2   Example of graphic representation used.

BIOFUELS: INDUSTRIAL, AGRICULTURAL AND ENERGY POLICIES[1]

Every sector is affected by a variety of policies. Since biofuels have most of their costs associated with raw materials of agricultural or forestry origin, they clearly relate to agricultural policy. Energy policy and regulation are also essential. Industrial policy can be more or less important depending on the country context. Policy coordination is critical for the early advancement of innovation and beyond.

Industrial, Agricultural and Energy Policies in the United States

The United States initiated policies for the use of ethanol as a vehicle fuel in order to ensure energy security in the 1970s during the first oil shock. In 1978, the National Energy Act benefited producers of gasohol (mixture of 10% ethanol with 90% gasoline) with tax credits. A fact not always explored is that at this time the United States had a surplus in grain production, and farmers and politicians from Midwestern states began to advocate for the implementation of national policies in support of ethanol. One notes here a junction of energy and agricultural policies in the direction of enhanced ethanol production and consumption.

With the advent of lower oil prices in the 1980s there was a significant change in interest in ethanol, prompting a decrease in resources for research and a shift of interest to longer maturity and high-risk projects. R&D efforts began to be directed to cellulosic ethanol using the enzymatic process (Wyman, 2001). In 1992, led by environmental concerns, the Environmental Protection Agency

---

1. This section concentrates on federal policy in both countries, thus abstracting from important forms of relevant industrial policy at the state level. For sure states are really important both in Brazil (mainly because of state tax "ICMS" varying between 12% and 30% which may make ethanol competitive or not) and in the United States because of grants and state regulations. An important reason for this limitation is shortage of space. Moreover, federal governments play instrumental roles. For instance, the Department of Energy arguably has a lot of instruments to provide incentives for innovation and assisting energy-related industrial fields.

(EPA) started the fuel oxygenation program requiring that 39 metropolitan areas blend ethanol or MTBE (methyl *tert*-butyl ether) to gasoline. In 1999 California banned the use of MTBE, regarded as a significant risk to the environment, leading to a general ban across the country. After this, ethanol was adopted as oxygenate blended with gasoline as a mandatory measure.

While the United States lacks a Department of Industry, a number of other government agencies engage in mission-oriented activities whose outcomes essentially amount to those of industrial and innovation policies. Despite not being the driver of the Renewable Fuel Standard (RFS), the Department of Energy (DoE) makes great efforts working closely with industry to ensure the availability of ethanol as well as ethanol's technological advancement. The DoE coordinates an extensive network of national research laboratories with extensive research, many of them founded decades back after World War II and maintained to the present day as the backbone of public R&D (U.S. DOE, 2014b).

DOE's Bioenergy Technologies Office (BETO), under the Office of Energy Efficiency and Renewable Energy (EERE), is focused on building partnerships and sharing the cost with key stakeholders in order to develop, demonstrate and deploy technologies for production of advanced biofuels from lignocellulose biomass and algae (U.S. DOE, 2014a). The strategic objective of BETO is to develop commercially viable technologies to use biomass to allow the sustainable production of biofuels at the national level that are compatible with today's transport infrastructure and allow replacing a portion of petroleum-derived fuels to reduce US dependence on oil and encourage the creation of a national bioenergy industry.

The National Bioenergy Center was established in Oct. 2000 at the National Renewable Energy Laboratory (NREL) to support the scientific and technological objectives of BETO, with a mission to promote the ability to catalyze the replacement of oil by transport fuels from biomass, offering innovative solutions of low-cost biofuels (NREL, 2014). The Centre is responsible for coordinating the efforts of several other DOE laboratories: (1) R&D biorefinery (NREL); (2) feedstock development (Oak Ridge National Laboratory); (3) biomass harvesting technology (Idaho National Laboratory); (4) syngas, catalysis and bioproducts (Pacific Northwest National Laboratory); and (5) reaction engineering and separations (Argonne National Laboratory).

In terms of agriculture, it is important to note that ethanol is mainly produced by corn starch in the United States, providing alternative market opportunities to corn producers, thus underwriting more stable demand and arguably increased financial gains. Moreover, cellulosic ethanol adds the opportunity for farmers to transform what is currently agricultural waste to a source of income. It also adds the possibility of including new crops in areas currently inhospitable to corn and with a huge market potential. Likewise, the US Department of Agriculture (USDA) also has a strong involvement in the development of biofuels, having established several programs to support the development of technology and to promote the inclusion of new energy crops.

Overall, it is observed that there is alignment between agricultural and energy policies regarding biofuels in the United States. While there is no explicit coordination, the size of the system creates impacts that resemble those of policy coordination and what in other countries would be described as the promotion of industry (rather than just technology). Yet, despite the large number of supported projects, results have not met the proposed demand for cellulosic and advanced biofuels. This led to the set up of an Interagency Working Group (IWG) on Biofuels with the participation of DOE, USDA and EPA and a central management team from the White House to oversee the coordination of efforts between the IWG and the Biomass Research and Development Board established by the Biomass Research and Development Act of 2000 and reauthorized in the Agricultural Act of 2014.

Industrial, Agricultural and Energy Policies in Brazil

In Brazil, the beginning of the mix of ethanol and gasoline for automobile fuel goes back to the 1930s and it is the result of agricultural policy. At the time there was a worldwide glut of sugar as a result of which sugar mills and cane growers were lobbying to expand the use of ethanol as an alternative use of sugarcane. Associated with the widening Brazilian trade imbalance due to growing gasoline imports, the situation of excessive supply of sugarcane led to the introduction of the ethanol-gasoline blend in Brazil. In contrast to corn-based ethanol, sugarcane provides a significant advantage regarding energy balance (the energy output/input ratio) and environmental performance. Although numbers can vary depending on assumptions regarding sugarcane productivity, transportation distances and electricity surpluses generated from its bagasse, according to Macedo et al. (2008) the energy ratio was 9.3 for the 2005/2006 harvest and could reach 11.6 in 2020 with new technologies. Goldemberg et al. (2008) show that these numbers can be five times higher than corn-based ethanol.

But, similarly to the United States, in modern times it was the first oil shock that increased interest in ethanol. In the 1970s Brazil was at the peak of its industrial policy of import substitution. The need to ensure energy security in the country gained tremendous momentum with this type of industrial policy already in place. At the same time, there was stagnation in exports of sugar which guaranteed support of the agricultural sector to creating new sources of demand. It is noteworthy that the ProAlcool policy had most actions carried out under the Industrial Technology Department of the Brazilian Ministry of Industry, making industrial and technology policy the main drivers of growth in the ethanol industry.

The elimination of the Ministry of Development, Industry and Trade in 1990, together with the consequent closing of the Industrial Technology Department that was responsible for conducting several technological development activities related to ProAlcool represents the end of the driving phase of the ethanol sector mainly by industrial policy. At this time ethanol started to be considered an agricultural product and be associated mainly with the Ministry of Agriculture, Livestock and Food Supply.

The Brazilian agency Embrapa is one of the largest agricultural research institutions in the world. In 2006 Embrapa Agro-Energy was established as a decentralized unit of Embrapa with the following goals:

1. Coordinate research activities in agro-energy;
2. Create a center of knowledge assemblage and specific expertise previously scattered across various units of the organization;
3. Hire specialists with knowledge not yet incorporated or internalized in the technical-scientific profile of the organization, but necessary to support the National Agro-Energy Plan;
4. Be recognized as a reference center from which Embrapa will integrate the networks and multi-institutional consortia formed for relevant research and innovation. Embrapa Agro-Energy is the result of efforts led by the agricultural policy to modernize the biofuel production technology.

In 2005, the Ministry of Science and Technology found that to meet the demand of fuels for the growing vehicle fleet, it would be necessary to overcome significant technological bottlenecks. In Jan. 2010 the National Laboratory of Science and Technology of Bioethanol (CTBE) was inaugurated in Campinas, whose stated goal is to study the cycle of sugarcane/ethanol, focusing on industrial technology of cellulosic ethanol. The aim is also to implement the no-tillage system in sugarcane cultivation and create sustainability models for the sector.

The publication of Provisional Measure 532 in Apr. 2011 initiated a phase in which the ethanol sector is being driven mainly by energy policy. Unlike the United States, where environmental policy leads the push for increased use of ethanol, environmental policy takes a backstage in Brazil seemingly considering that the ethanol industry is already deployed and that it does not require additional actions to grow.

The same year (2011) BNDES (Brazilian Development Bank) and FINEP (Brazilian Innovation Agency) launched the Joint Plan of Support for Technological Innovation of the Industrial and Chemical Industries using sugarcane (PAISS), based on the diagnosis of Nyko et al. (2010) that in Brazil there was: (1) poor coordination between stakeholders in the sector (companies, research organizations and financial organizations); (2) low participation of the private sector in innovation investments; and (3) insufficient, diffused and unfocused incentives. The outcome of PAISS in 2015 is the establishment of two industrial and one demonstration plant producing E2G.[2]

---

2. Second-generation *ethanol* can be produced from various types of biomass, such as lignocellulosic biomass, woody crops, agricultural residues, waste or algae. First-generation ethanol is made from sugars.

Nyko et al. (2013) found that the agricultural sector had reduced the rate of creation of sugarcane varieties and that there was a low degree of development of agricultural machinery for the production of ethanol. In Feb. 2014, a joint action of BNDES and FINEP released the Agricultural PAISS aiming to foster both the development and pioneering production of agricultural technologies such as the adaptation of industrial systems as long as they are inserted into the supply chains of sugarcane and/or other energy crops that are compatible and/or complementary to the sugarcane agro-industrial system (Brazil, 2014).

In recent years significant actions have been implemented to foster technological development associated with the production of ethanol, both first-generation (1G) and second-generation (2G). New technologies in the energy sector normally compete with technologies already well established, with low costs and large installed base, so that the biggest problem faced by the innovations in the sector is to achieve the necessary scale. In the energy sector it is necessary that the TIS takes into account technology supply, but also the creation of demand for this type of technology and organizational elements that are properly aligned to connect supply and demand (Peters et al., 2012; Weiss and Bonvillian, 2009; Ye et al., 2014).

The lack of specific demand for E2G shows that energy policy has not worked in alignment with the industrial and technological policies in Brazil. As California pays a premium value for E2G, and EPA promotes its use, the Brazilian production of second-generation ethanol is mainly exported to the United States.[3]

ETHANOL THROUGH AN ENVIRONMENTAL PERSPECTIVE

Environmental Impacts of Fuels

Energy production and consumption entail a set of local and global environmental impacts that increasingly influence public opinion and public policies. The national energy policy, for instance, may aim to change the composition of the energy mix in order to minimize these environmental impacts by promoting the substitution of fossil fuels with renewable energy sources, with lower negative impact on global climate change.

Fossil fuels such as oil, coal and natural gas are at the heart of the debate on climate change because burning hydrocarbons releases significant amounts of carbon dioxide (CO_2) to the atmosphere, the main gas responsible for the greenhouse effect. These three fossil fuels collectively account for 85% of all anthropogenic greenhouse gas (GHG) emissions (IPCC, 2007).

---

3. It should be mentioned here that there is no difference in the final product (1G and 2G ethanol). The production method is very different and the environmental results are very different too. Depending on the production method, E2G will reduce the CO_2 emission much more than E1G. In some cases such as E2G using energy cane there is no emission and the complete cycle capture carbon. Consumers do not care. The environmental agencies prefer E2G.

The known reserves of fossil fuels have a carbon content estimated at 3863 billion tons (Gt) of CO_2, while the "carbon budget" to ensure, with a probability of 66%, that the increase in global temperature is contained within the 2°C limit is 1050 Gt of CO_2 (IPCC, 2013). Therefore, the key challenge for climate change management is changing the global energy matrix from the current dominance of fossil fuels to less carbon-intensive energy sources. In addition to carbon dioxide (CO_2), burning fossil fuels increases the production of particulate material and the flow of atmospheric nitrogen and sulfur oxides (SO_2, NO and NO_2) and carbon monoxide (CO).

Environmental impacts, however, are not restricted to fossil fuels. All energy sources, even renewable energy, carry some environmental impact. Often, however, these impacts occur at different stages of the energy life cycle. Proponents of specific technologies typically will downplay the environmental impacts of their favorite technology while stressing the impacts of competitors in order to gain legitimacy.

The environmental costs of energy production and consumption must be taken into account in the public policies for the sector, especially since most of these impacts are considered an externality by energy producers and are not reflected in the price of energy products. This is, of course, the classic negative externality effect in economics (social cost larger than private cost) that causes production above the socially optimum level. In addition, the strong correlation between energy consumption and human development has justified government subsidies to traditional fossil fuels, making the decarbonization of the global energy mix even more challenging.

The New Energy Paradigm and the Fuel Mix

The pre-eminence of global climate change on the international agenda has brought a new set of political imperatives that call into question the conditions of production and consumption of fuel, affecting the basic premises around which energy policy was built. The World Energy Council, for example, asserts the existence of a "trilemma" between energy security, equitable access to energy and environmental sustainability.[4] Analysts also speak of the emergence of a "New Energy Paradigm" (Helm, 2007) or of the occurrence of a "New Energy Crisis" (Chevalier and Geoffron, 2013).

In this new paradigm, energy policy, which since World War II has been considered a strictly national issue, attains a global dimension. Climate change has introduced in this context a new global public good: climate protection ensured by vigorous mitigation of emissions of GHGs. The incorporation of climate protection in energy policy implies not only its direct connection to the foreign

---

4. See chapter "Political Orientations, State Regulation and Biofuels in the Context of the Food–Energy–Climate Change Trilemma."

policy agenda but also an opportunity for exercising international leadership, taking as a criterion the ability to promote GHG emission reduction.

Due to the importance of energy security to economic growth and development, a contemporary trend in energy policy is the search for flexibility in the fuel matrix with diversification and increasing participation of biofuels, reduction of dependency on fossil fuels, and lessening of energy vulnerability, that is, reducing the need for energy imports and increasing the number of suppliers. Such flexibility involves the inclusion of energy alternatives that, in addition to seeking security of supply, intends to bypass the likely depletion of cheap sources of fossil resources and can also contribute to reducing the adverse effects of conventional energy consumption on global warming (Kuzemko, 2013).

In recent decades, the introduction of flex-fuel vehicles and the start of compulsory blending of biodiesel into mineral diesel have enabled the biofuels sector to gain a considerable share of the Brazilian energy matrix. Ethanol assumes central importance because while its use was encouraged in response to an external constraint of fossil fuels, it has also contributed significantly to reducing GHG emissions in the transportation sector. The Ministry of Science and Technology estimates that from 1990 to 2008 the substitution of gasoline for ethanol in flex-fuel cars contributed to an emission reduction of 13 million tons of CO_2 per year (Brasil, 2007). The positive energy balance and constant improvement in sustainable agricultural practices in the sector have given sugarcane ethanol the status of a climate-friendly substitute to fossil fuels supporting the legitimization function of the ethanol TIS.

The Brazilian Intended Nationally Determined Contribution to the UNFCCC, in preparation for the Paris Summit, commits the country to raising the share of sustainable biofuels in the energy mix to approximately 18% by 2030 by expanding biofuel consumption and increasing ethanol supply, including increasing the share of advanced biofuels (E2G) and increasing the share of biodiesel in the diesel mix.[5] This international commitment underlines the role of biofuels in the political economy of global decarbonization and signals the political will to enhance national policies for biofuels.

The expected extensive gains in productivity and the possibility of an even more favorable GHG emission balance promises an increasingly relevant role for biofuels for decarbonization of the fuel and energy mix (Souza et al., 2015). Also, the unique possibility of negative emission by means of a combination of biomass energy with carbon dioxide capture and storage (BioCCS) further supports the legitimization function. A study commissioned by the IEA Greenhouse Gas R&D Programme, for instance, found an annual global potential of up to 10 Gt of negative CO_2 emissions in the year 2050 (IEAGHG, 2011).

---

5. Available from: http://www4.unfccc.int/submissions/INDC/Published%20Documents/Brazil/1/
BRAZIL%20iNDC%20english%20FINAL.pdf.

BIOFUEL REGULATION

Although the development of alternative energy sources can be partially attributed to energy security issues—aiming at reducing imported petroleum—more recently environmental criteria have become the main drivers of biofuels production. While environmental and energy policies have their distinct characteristics that prevent their conflation (Ellerman, 2012), biofuels and land use provide an important link. This section discusses the introduction of this public policy feature in the two countries, focusing primarily on specific regulatory aspects that can foster or hinder innovation in advanced biofuels.

Regulation in the United States

The starting point to understanding biofuels regulation in the United States is the RFS. Created under the Energy Policy Act of 2005, RFS is one of the main federal programs for biofuel production support whereby an annual minimum volume requirement is set for transportation fuel content.

At first, this program required 7.5 billion gallons of renewable fuel to be blended into gasoline by 2012. In the following years, it included diesel and determined a schedule to gradually increase year by year the minimum requirement for the transportation fuel supply to reach 36 billion gallons, nearly a fivefold increase, by 2022. The most important change was the establishment of separate mandates according to the categories of renewable fuels, advanced biofuels, biomass-based diesel and cellulosic biofuels. These categories rely on each biofuel's potential GHG emissions reduction compared to the baseline fossil fuel (gasoline or diesel) that the biofuel replaces.

EPA uses specific models for lifecycle GHG emission measurement for each biofuel pathway. A fuel pathway distinguishes fuel types by feedstock[6] used (eg, energy cane, giant reed, camelina) or production process (eg, thermochemical pyrolysis, esterification). It is assessed according to a lifecycle assessment that includes not only direct emissions from the use of the final fuel, but also significant indirect emissions such as those from land use changes for feedstock production and fuel distribution.

Ethanol made from corn starch only qualifies as renewable fuel in this framework, while ethanol made from Brazilian sugarcane qualifies as advanced biofuel due to its potential GHG emissions reduction. These differences are important in terms of minimum requirement, which is translated in a subsidy resulting from the sale of Renewable Identification Numbers, the economic instrument based on a tracking system through which mandates are accomplished in a flexible manner by refiners, blenders and importers across the country.[7]

---

6. All renewable fuel must be made from feedstock that meets specific requirements to be qualified as renewable biomass, including land use restrictions.
7. Details of the framework for implementation of a market for RINs under the RFS are presented in Schnepf and Yacobucci (2013).

The RFS intended to create a mandatory market for renewable fuels, including ethanol, in order to reduce the risk associated with the capital-intensive structure that characterizes second-generation biofuels. However, there are practical constraints on the supply of higher ethanol blends to the vehicles that can use them (the so-called "blend wall") which is forcing EPA to reconsider initially set targets and delay announcements on the minimum volumes for subsequent years.[8] This restriction comes from the Clean Air Act (CAA) regulatory framework for new fuels and additives, the second reference for understanding the regulation of biofuels in the United States.

Under the CAA, the Environmental Protection Agency (EPA) is the sole authority in the United States to regulate lawful commercialization of new fuels and additives. Specifically, the CAA requires that new fuels are "substantially similar" to those already used in certification processes for engine emissions controls systems and the EPA provides detailed specifications that end-fuels must meet to be considered as such. To overcome this restriction, biofuel producers can resort to the "substantially similar" requirement and request a fuel waiver or registration of a concentration that complies with an existing waiver issued by the EPA. Slating and Kesan (2011) point out that these regulatory paths to reach lawful commercialization can be very burdensome to proponents of new fuels and additives, highlighting the need for regulatory innovation to provide more clear signals of market opportunities for advanced biofuels.

In addition, advanced biofuel producers face an important uncertainty related to the maintenance of targets initially set for the RFS, since the EPA has the authority to waive the statutory volume requirements, in whole or in part, if the agency determines that their implementation "would severely harm the economy or environment of a State, region, or the United States, or there is an inadequate domestic supply."[9]

Regulation in Brazil

Notwithstanding Brazil's long tradition in the production of ethanol from sugarcane, the regulatory framework for biofuels evolved and gained prominence only recently when the country started to produce biodiesel from different feedstock. Currently, the body responsible for regulating the production,

---

8. "Relying on its Clean Air Act waiver authorities, EPA is proposing to adjust the applicable volumes of advanced biofuel and total renewable fuel to address projected availability of qualifying renewable fuels and limitations in the volume of ethanol that can be consumed in gasoline given practical constraints on the supply of higher ethanol blends to the vehicles that can use them and other limits on ethanol blend levels in gasoline. These adjustments are intended to put the program on a manageable trajectory while supporting growth in renewable fuels over time" (EPA, 2014, p. 71732).
9. U.S. Code § 7545—Regulation of fuels (o)(7)(A)(i) and (ii).

distribution and commercialization of biofuels in Brazil is the National Agency for Petroleum, Natural Gas and Biofuels (ANP), a special agency subordinated to the Ministry of Mines and Energy (MME) which is in charge of implementing energy policies.

In the downstream segment, ANP's competence relates to ensuring the supply of petroleum products and biofuels nationwide, and establishing their technical specification. From 2005 on, ANP's main goal regarding biofuel regulation has been to monitor fuel quality to prevent fraud. Only recently, the agency has started to authorize biofuel tests for assessment of nonspecified blends or of new fuels to be used in captive fleet or industry.

The regulatory framework for biodiesel was established with Federal Law No. 11.097/2005, while ethanol, no longer regarded an agricultural product, became an energy product with the publication of Provisional Measure 532 in Apr. 2011, representing a challenge to the agency created to regulate the oil and natural gas markets due to the different market structure. Law No. 12,490 of 16/09/2011 amended Law No. 9,478/97 extending the competence of the ANP for all biofuels. The introduction of markets that depend on agricultural production, dispersed among a large number of private players, further increases the need for coordination of regulation with sectoral public policies.

While the RFS in the United States sets progressive targets for renewable fuels in the timeframe 2010–22, the Brazilian market for biofuels is guaranteed by minimum requirements of ethanol in gasoline and finds important support in the increasing flex-fuel fleet with vehicles that run on pure ethanol. Recent R&D projects induced by sugarcane productivity have aimed at introducing innovations in the production process, thus presenting potential implications for the regulatory authorities.

COMPARISON OF INNOVATION FUNCTIONS BETWEEN THE UNITED STATES AND BRAZILIAN BIOFUEL TIS

Feedstock

Due to the use of sugarcane as raw material, the Brazilian 1G ethanol already presents a considerable reduction in GHG emissions and faces little environmental opposition. Besides, the food versus fuel debate did not gain as much relevance in the country as it has in the United States, implying that the association of higher ethanol production with a possible reduction in food production is hardly an issue among the public. Ethanol is viewed sympathetically by the population and is widely used as long as it has a competitive price. The flex-fuel car tax incentives have been a big success, creating the potential for huge demand for ethanol. All these conditions make up a *well formed market (function 5)* and go without the creation of advocacy coalitions for an *expansion of market for 2G ethanol (function 7)*.

In the United States, the fact that corn starch ethanol accounts for a very small GHG reduction spurred several criticisms about the use of biofuel. The presence of a wide range of interest groups that promoted the food versus fuel debate associating biofuels with low environmental performance has also created opposition to the use of corn starch ethanol. The intense opposition for first-generation (1G) ethanol, coupled with the environmental focus of the policy, led to massive investment in second-generation (2G) ethanol and the creation of a mandatory demand.

In Brazil the main obstacle to the expansion of ethanol is its competitiveness in terms of price. Ethanol has not been able to match the price of gasoline except in states that are producers and have beneficial tax policies such as reducing the state tax (ICMS). Being widely used in major producing regions (São Paulo, Goiás, Mato Grosso, Paraná and Minas Gerais), ethanol represents 50% of the Otto cycle vehicle fuel in the country (May 2015), in contrast with the severe limitations posed by the use of corn starch ethanol and by the "blend wall" in the United States.[10]

As there is little opposition to 1G ethanol in Brazil, there is limited action to develop new technologies or to encourage this market. Although the result of coordination between FINEP and BNDES, the PAISS is not a permanent initiative for the support of 2G ethanol and new products based on sugarcane. Despite having resulted in the installation of two cellulosic ethanol plants and an additional project being implemented, there is no market reserve for their product. The implication is that the competitiveness of 2G ethanol will be exclusively price-based. Despite the fact that the 2011 law contains a modern definition of ethanol, the regulation of ANP does not differentiate cellulosic ethanol. There is no market reserved specifically for cellulosic ethanol in the country: the technology's success rests solely on price relative to 1G ethanol, a significant barrier for a technology in its infancy.

Conversely, the United States counts on a much more developed R&D structure that forms an important basis for the *knowledge development (function 2)* and testing technology applications for biofuels production. The capacity of this structure to form networks must be highlighted where information can be shared across a large number of actors, presenting strong *knowledge diffusion through networks (function 3)*.

---

10. The "blend wall" is the maximum quantity of ethanol that can be sold each year given legal or practical constraints on how much can be blended into each gallon of motor fuel. The Renewable Fuels Association claims the sole reason for its existence is that the oil industry has refused to invest in blender pumps, storage tanks and other infrastructure compatible with E15-and-higher ethanol blends (http://www.ethanolrfa.org/issues/blend-wall/).

Maintaining a Long-Term Commitment

From a regulatory perspective, a few comments can be made to compare the two countries. The first is that the mandate-based policies for biofuels in both countries generate market uncertainty. In the United States uncertainty arises from the possibility of the EPA cutting off initially announced targets, as happened in late 2013 to RFS2 targets due to constraints on the supply of higher ethanol blends in gasoline, and limits on the availability of advanced biofuels. In Brazil, mandates are set mainly based on supply capacity, with other factors being potentially relevant depending on political and/or economic circumstances. However, in recent years inflation control has played an important role in fuel pricing management, directly affecting the market for ethanol. Much less certainty is offered to ethanol producers by the Ten-Year Plan for Energy Expansion (PDE),[11] which does not put forward a long-term commitment regarding the supply of ethanol and other sugarcane products (Table 4.3).

US policy requires a considerable expansion of biofuel production and consumption in the upcoming years, which will only be possible with the active participation of the private sector. Even though the public sector is not supposed to deliver biofuels to the market, it can set incentivizing rules for the private sector actors to do so. Under the Brazilian regulatory framework there is little

TABLE 4.3 Provisions for the Sugarcane Sector in the Plan for Energy Expansion

	Hydrated Ethanol (Billion Liters)	Anhydrous Ethanol (Billion Liters)	Ethanol Total (Billion Liters)	Final Energy Consumption of Sugarcane Bagasse (10^3 toe)
PDE 2022	36	18	54.5	43,438
PDE 2021	51.98	9.63	61.61	50,010
PDE 2020	55.88	7.18	63.06	50,698
PDE 2019	47.25	5.12	52.37	53,466
PDE 2017	48.77	4.44	53.21	22,162
PDE 2016	22.62	6.63	29.25	21,736

11. The PDE is the Brazilian medium-term energy planning instrument and presents important signals to guide the actions and decisions related to the balance of energy supply and demand based on 10-year projections. The Energy Research Enterprise, which is linked to the Ministry of Mines and Energy, is responsible for its preparation.

incentive to develop advanced biofuels arising from demand side measures, as neither final consumers nor fuel distributors can distinguish the value of 2G over 1G ethanol (besides the price). These facts reveal the need for credible policy goals over time.

For this reason, Kay and Ackrill (2012) defined five capacity categories to assess the regulator's ability to deal with long-term goal commitment. Interestingly, they somehow find correspondence with the SFs either directly or indirectly. The first category refers to the capacity of being supported by a broad-based coalition, even though for sustainability issues it is very difficult to introduce important policy changes without facing a strong and adverse response from certain individuals and interest groups. The second category comprises the selection of policy alternatives and technologies that should be prompted to reach the set goals. Third is the capacity to work across different policy networks to deal with political, technological and market uncertainties and complexity in a broad sense and operationalize defined policies. The last two categories are of central relevance to biofuel policymaking because they aim to address explicitly uncertainty issues. Foresight capacity demonstrates governments' ability to look forward instead of being led primarily by short-term electoral cycle myopia. Reflection capacity reveals the ability to monitor on a continuous basis whether the long-term goals will be reached with a high degree of confidence.

These capacities relate to a particular characteristic of the US regulation, expressed by the temporal separation of 1G and advanced biofuels targets that has allowed the government to implement the regulation in spite of market and technological uncertainties placed by the blend wall and by the potential barriers for advanced biofuels scale up. In this regard, the US strategy has been to "agree policy objectives in the short term, and accommodate different and conflicting values involved in biofuels expansion by legislating guides to future policy change in the advanced biofuels sector" (Kay and Ackrill, 2012, p. 301).

The way the government announces its targets can be decisive for defining the biofuels sector trajectory as they can impair the much needed long-term certainty regarding the existence of markets for advanced biofuel producers. Overall, despite the existence of market uncertainty in both cases, the US framework provides much clearer signals for advanced biofuel technologies than the Brazilian framework, which does not provide demand side incentives for such technologies. Furthermore, the considerably longer timeframe in US regulation provides stronger signals of intertemporal commitment to biofuel support than the Brazilian regulation.

Setting an Adequate Environment for Innovation

Under the RFS, there is limited guidance on technologies, since the supporting policies are meant to be "technology-neutral" (Kay and Ackrill, 2012). This means that having the goals set, the private sector and other nonstate actors compete on an equal basis to find the best solutions. Second-generation biofuels are part of the so-called industrial biotechnology for which R&D activities are highly important. As such, they are subject to regulation of other types of activities, such as research related to genetically modified organisms (GMOs), which makes the institutional design even more complex. In the United States there are three main agencies involved in regulating GMOs: the US Department of Agriculture's Animal and Plant Health Inspection Service (APHIS), the Food and Drug Administration (FDA) and the Environmental Protection Agency (EPA). Despite the higher institutional complexity that stems from the interaction of more regulating agencies, representatives from the productive sector affirm they have preferred to carry out biotechnology R&D in the United States due to more expeditious processes compared to the Brazilian framework (de Sousa, 2015).

It is worth noting that when firms are subject to overlapping regulatory jurisdictions, they will try to reap opportunities that arise from gaps between the jurisdictions of regulatory agencies, by being subject to regulation under which they will have better outcomes. This can particularly be the case of the regulation of GMOs for R&D activities that support biofuels innovation.

While industrial biotechnology in Brazil finds potential in terms of biodiversity and competitiveness in biomass production, the country may also present conditions for innovative firms to develop R&D activities in other elements, resembling a situation that Joskow (2010) defines as "regulatory arbitrage." This phenomenon exists "between federal and state regulatory agencies, between federal regulatory agencies with overlapping jurisdictions, and, when there are opportunities to exploit differences in regulatory frameworks, in different countries" (Joskow, 2010, p. 6). Another regulatory issue that can make innovating companies migrate to other jurisdictions is the delay in authorizing new fuels or new fuel blends (*regulatory delay*). Prieger (2007, p. 220) states that it occurs when the regulation "does not allow the introduction of new products without regulatory review or approval." There is evidence that increased delay is associated with fewer new products introduced.

Both examined regulatory frameworks present considerable regulatory delay for the introduction of new fuels. Both the EPA and the ANP require public notice and opportunity for comment before determination on significant changes regarding technical specification and waiver of requirements. Interestingly, the fact that the regulator must review and approve the commercialization of new products should be considered good practice, since it eventually prevents society from being in contact with harmful practices, products or substances. Regulatory delay can have additional beneficial effects such as allowing for the resolution of coordination or technical standards issues, which can reduce costs from the supply side or increase demand (Prieger, 2007).

As 2G ethanol presents an evolution of production process compared to 1G ethanol and no change in the chemical characteristics of the final product, it may face few obstacles in terms of technical specification. However, the regulatory delay can be an important obstacle for other renewable products, considering that 2G biofuel technologies are often being developed in a new biobased industry paradigm, so-called biorefineries. Biorefineries enlarge the product spectrum, and hence the market possibilities, for companies investing in products based on biomass as a main raw material through different ways of innovation described in detail by Bomtempo and Alves (2014).

Institutional Design's Influence on Proactivity

A final remark must be made by distinguishing both regulatory frameworks through the prism of institutional design. First, it must be emphasized that EPA's primary duty is environmental regulation. The agency was created to consolidate a wide variety of activities to ensure environmental protection, including research, monitoring, standard-setting and enforcement. The GHG emissions regulation was recently assigned to EPA. The ANP, on the other hand, was created to regulate the oil and gas industry, expanding its regulatory jurisdiction to biofuels only recently. Naturally the Brazilian agency has focused on biofuel technical specifications and quality control, leaving aside technological innovation concerns in its production. The differences in the primary duties of the competent agencies explain to a great extent EPA's proactive role in determining ambitious targets for advanced biofuels and in providing signals for long-term investments through the RFS, in contrast to ANP's performance characterized by more passive regulation.

Specific mandates demand more institutional capacity from the EPA to better monitor the evolution of liquid fuels markets and set the regulatory framework to foster innovation and enforce minimum requirements. Regarding the technical capacity to develop and manage a diverse number of fuel pathways, the EPA recognizes that "lifecycle GHG assessment of biofuels is an evolving discipline" and will continue to revisit their analyses in the future as new information becomes available.[12]

The shale gas revolution in the United States has reduced the weight of energy security as an argument for biofuels support. In Brazil, a similar result could be observed due to the large pre-salt oil reserve discoveries. In this context, biofuel pathways certification will become increasingly important as it ensures that sustainability criteria are met and contribute to justify government support and to differentiate final fuels to the average consumer who currently would see no value in 2G ethanol.

Two extensive maps of the Technology Innovation System of Ethanol (TISE) in the United States and Brazil have been drawn on the basis of the policy and regulation aspects described above (Figs. 4.3 and 4.4). Rather than comprehensive, the maps summarize the most important aspects of the TISE in the two countries.

---

12. Environmental Protection Agency 40 CFR Part 80 Regulation of Fuels and Fuel Additives: Changes to Renewable Fuel Standard Program. Federal Register/Vol. 75, N. 56/March, 26, 2010/Rules and Regulations, p. 14670.

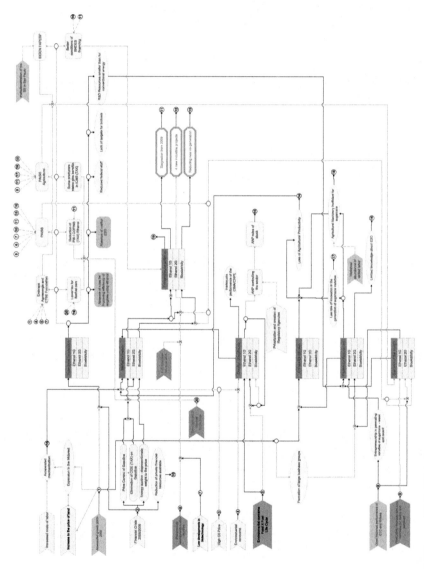

FIGURE 4.3 TISE maps (United States).

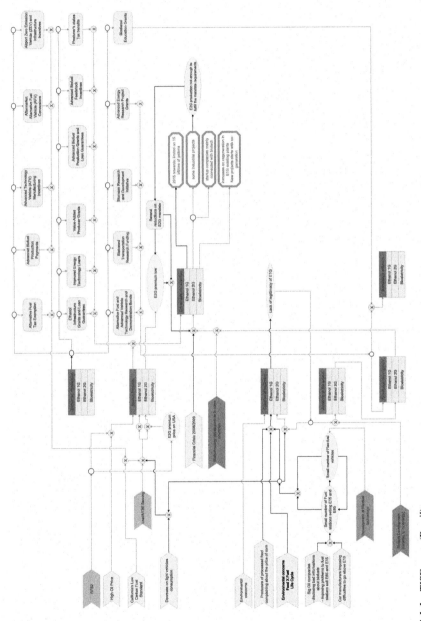

FIGURE 4.4 TISE maps (Brazil).

Table 4.4 shows a comparison between the two innovation SFs. In Brazil, the function that currently is judged to have the worst performance is the knowledge development function. In contrast, the market formation function is the one with better performance and has "pulled" the development of the TISE, even though this function had been negatively affected in the period 2008–14 by the gasoline price containment policy of the federal government.[13] In the United States, the legitimization function is the one with the worst performance and has in turn negatively affected the market formation function. In contrast, the function of knowledge development stands out quite strongly. It must be noted here that numerous actions related to the creation of knowledge can improve legitimacy. As advanced biofuels have the potential to increase fuel production by using agricultural residues (and not noble feedstocks), the development of E2G may be included as an action that responds to the food versus fuel debate.

Another very important difference between the two countries is policy coordination. In the United States a strong coordination of actions, mainly matching demand generation with investment in technology, has been pointed out. In Brazil, there is no similar connection between the efforts of new technology development and demand generation. Moreover, the Interministerial Council for Sugar and Alcohol (CIMA) which would have the task of coordinating TISE government actions is ineffective (Sousa and de Carvalho, 2013).

FINAL REMARKS

Brazil and the United States account for over three-quarters of the world production and consumption of ethanol. This is due to the two countries having large fleets of motor vehicles, very large agricultural areas, efficient agricultural production systems and multifaceted public policies for the sector. The evolution of ethanol use worldwide is strongly linked to how innovation will evolve in these two countries in the near to medium future. In this chapter, we use the TIS approach to compare elements that can foster or prevent the wider deployment of the so-called second-generation ethanol (E2G) and, more generally, the expansion of ethanol's share of the transportation fuel mix.[14]

Brazil has a better structured market for ethanol use, reflecting the positive impact of past investments in the sector. However, currently the system shows significant inertia that prevents wide-scale adoption of the next generation of production technology (E2G). The United States has difficulties with market formation, but has taken a well-articulated set of actions to develop technology that can significantly alter this picture.

---

13. Petrobras, Brazil's main supplier of fuel, has been restricted (by the federal government) to increase fuel prices in order to control inflation.
14. This chapter has not addressed extensively the very important issue of biofuel sustainability. For extended recent treatments see Souza et al. (2015) and Elbehri et al. (2013).

TABLE 4.4 Comparing TISE Functions Between Brazil and the United States

	Brazil	United States
Entrepreneurial activities	E1G—Business responds when the price is adequate and there is a large market E2G—PAISS plus US premium price have stimulated some business capital investment in E2G. It is limited to businesses connected to PAISS projects (almost no venture capital)	E1G—Business responds when the price is adequate, but they face the "blend wall" barrier E2G—RFS mandate + US premium price + DOE and USDA grants have stimulated business to E2G – intense creation of startups mainly connected with biotech (venture capital plays a relevant role)
Knowledge development	E1G—Growing less than price of other production factors E2G—Mainly imported technology Low biotech development	E1G—Growing E2G—High development High biotech development
Knowledge diffusion	E1G—2003–09 high rates of growth created problems	A lot of events and resources to spread the technology
Search guidance	Basic R&D focused on E1G PAISS stimulates E2G	Government focus on E2G A lot of private R&D to improve yield of corn ethanol
Market formation	Flex-fuel vehicles sales growing fast All fuel stations have pumps to sell pure ethanol Consumption of light vehicles growing fast Control of gasoline price No E2G market	Just few flex-fuel vehicles Just few E85 gas stations Consumption of light vehicles decreasing Mandate to E2G Continuous reductions of E2G mandate
Resource mobilization	Small team working in the sector Fewer resources than traditional forms of energy PAISS PAISS Agricultural	Huge resource mobilization in terms of both personnel and financial commitment
Legitimacy creation	Almost no opposition	Huge opposition from big oil companies and car manufactures Debate food versus fuel Debate about GHG reductions

Although the knowledge creation function presents significant deficiencies in Brazil, other well-developed SFs still provide the conditions for E2G to be developed and widely adopted in the country. In particular, the potential for high-yield biomass production and efficient agricultural organization, coupled with a well-structured market provide a strong background for positive action in the Brazilian TISE.

E2G has huge potential to expand fuel supply without requiring land use expansion as scientists develop more productive varieties. This can be particularly important considering the implementation of a New Forest Code in Brazil—that imposes restrictions in expanding the agricultural area—and the intense food versus fuel debate in the United States. Milanez et al. (2015) show that the technological evolution of E2G, intercropped with a raw material of high productivity (sugarcane), has the potential to reduce the cost of ethanol to somewhere around US$45 per barrel of oil equivalent, making the biofuel competitive on price with fossil fuels.

Furthermore, depending on the technology and raw materials used, the ethanol plant may have an electricity surplus, as currently happens in Brazil with ethanol production from sugarcane. Coordination between the liquid fuels policy and the power to acquire that electricity surplus allows mitigating the risks of ethanol production, as the power purchase agreements are longer term and more stable in prices. This alternative also has environmental benefits in order to more efficiently use a raw material already extracted from nature that would need to be disposed into the environment or burned inefficiently.

Solecki et al. (2013) identified 93 companies in the United States and Canada providing inputs, catalysts, enzymes and technology services for the supply chain of advanced biofuels. According to these authors, between 2007 and the first half of 2013, there were US$1.45 billion of private investment in this segment. It is also reported that public grants since 2008 exceeded US$600 million and that the federal loan guarantees for biorefinery projects exceeded US$940 million.

A fact to note is that in the United States a significant part of technological development for biofuels is done by startups with technologies often found in the research laboratories of universities and developed with financial assistance from venture capital (Manzer, 2013). In Brazil, there are few startups and technology development is still dominated by the research institutes and large firms. A number of US startups have business in Brazil or have received investment from Brazilian companies (Solecki et al., 2013).

Although there are deep differences in terms of policy systems supporting innovation between the two countries, the results in terms of E2G industrial production are quite close. Brazil has two industrial units of large-scale operation and the United States has three. In Brazil there is a third E2G industrial unit specified to start construction. None of these units has managed to operate continuously. In both countries technical problems have resulted in frequent interruptions for repairs and redesign.

Many experts express optimism since the technical area that was considered the most uncertain (biotechnology) has performed satisfactorily. The main problem relates to the need to reduce costs. Ethanol producers and enzyme providers are saying that this is just a question of scale. If the production increases the costs will be lower. The main challenges in the Brazilian industrial units are related to the stage of pretreatment, focusing on mechanical problems of the equipment. Technicians consider that the problems are as expected and that there are no major technological challenges at this stage other than necessary engineering adjustments.

As biofuel production has most of its costs associated with raw materials, the strength of the agricultural and forestry sectors in the United States and Brazil, combined with the growing environmental concerns and with existing policies to support biofuels, may lead to the conclusion that both countries will be transitioning to E2G. Sustainability is a strong driver of technology innovation in ethanol production, helping lower costs over time. As Brazilian regulation does not set any target related to GHG emissions reduction, nor significant signals for market formation, E2G success depends on its capacity to be economically competitive. E2G insertion speed is largely dependent on government policies, in view of energy sector characteristics and in view of the initially higher price of this alternative. In Brazil, the lower intensity of knowledge development could lead to imports of technology and/or capital goods. Extant partnerships suggest that much of this technology will be imported from the United States and Canada.

ACKNOWLEDGMENTS

Nick Vonortas acknowledges the infrastructural support of his home unit, the Center for International Science and Technology Policy (CISTP), at the George Washington University. He acknowledges the generous support of FAPESP through the São Paulo Excellence Chair in technology and innovation policy at the University of Campinas (UNICAMP). He also acknowledges support from the Basic Research Program at the National Research University Higher School of Economics (HSE) within the framework of the subsidy granted to the HSE by the Government of the Russian Federation for the implementation of the Global Competitiveness Program. All authors wish to thank the book editors for very helpful comments on earlier drafts. Remaining mistakes and misconceptions are solely the responsibility of the authors.

REFERENCES

Bergek, A., et al., 2008. Analyzing the functional dynamics of technological innovation systems: A scheme of analysis. Res. Policy 37 (3), 407–429. abr.

Bomtempo, J.V., Alves, F.C., 2014. Innovation dynamics in the biobased industry. Chem. Biol. Technol. Agric. 1, 19.

Borup, M. et al., 2013. Indicators of Energy Innovation Systems and Their Dynamics. A Review of Current Practice and Research in the Field. Disponível em: <www.eis-all.dk>.

Brazil, 2014. BNDES. PAISS Agrícola. Retrieved from: <http://www.bndes.gov.br/SiteBNDES/bndes/bndes_pt/Areas_de_Atuacao/Inovacao/paissagricola.html> (12.05.14.).

Brasil, 2007. MCT. Contribuição do brasil para evitar a mudança do clima. Available from: <http://www.mct.gov.br/upd_blob/0018/18290.pdf>.

Chevalier, J.-M., Geoffron, P., 2013. The New Energy Crisis: Climate, Economics and Geopolitics, second ed. Palgrave Macmillan, London.

de Sousa, L.C., 2015. O setor sucroenergético e sua dinâmica de inovação (Doctoral disseratation), Universidade de Brasília, Brasília, Brazil. Retrieved from: <http://repositorio.unb.br/bitstream/10482/18490/1/2015_LucianoCunhadeSousa.pdf>.

Elbehri, A., Segerstedt, A., Liu, P., 2013. Biofuels and Sustainability. Food and Agriculture Organization of the United Nations (FAO), Rome.

Ellerman, D., 2012. Is conflating climate with energy policy a good idea? Econ. Energy Environ. Policy 1 (1).

EPA, Environmental Protection Agency, 2014. Standards for the Renewable Fuel Standard Program. Proposed Rule, 78 Federal Register No. 230, November 29, 2013.

Goldemberg, J., Coelho, S.T., Guardabassi, P., 2008. The sustainability of ethanol from sugarcane. Energy Policy 36, 2086–2097.

Hekkert, M.P., Negro, S.O., 2009. Functions of innovation systems as a framework to understand sustainable technological change: Empirical evidence for earlier claims. Technol. Forecast. Soc. Change 76 (4), 584–594.

Hekkert, M.P., Suurs, R.A.A., Negro, S.O., Kuhlmann, S., Smits, R.E.H.M., 2007. Functions of innovation systems: a new approach for analysing technological change. Technol. Forecast. Soc. Change 74, 413–432.

Helm, D., 2007. The New Energy Paradigm. Oxford University Press, Oxford.

IEAGHG, July 2011. Potential for Biomass and Carbon Dioxide Capture and Storage. 2011/06. Available from: <http://www.ieaghg.org/docs/General_Docs/Reports/2011-06.pdf>.

IPCC. 2007. Climate Change, 2007. Mitigation of Climate Change: Working Group III Contribution to the Fourth Assessment Report of the IPCC. Geneva.

IPCC, 2013. IPCC Fifth Assessment Report (AR5)—The Physical Science Basis. Geneva.

James E. Prieger, 2007. Regulatory delay and the timing of product innovation. Int. J. Ind. Organ. 25 (2), 219–236. ISSN 0167-7187, <http://dx.doi.org/10.1016/j.ijindorg.2006.05.001>.

Joskow, P., March 2010. Product market regulation: market imperfections versus regulatory imperfections. CESifo DICE Report 8 (3), 3–7.

Kay, A., Ackrill, R., 2012. Governing the transition to a biofuels economy in the US and EU: accommodating value conflicts, implementing uncertainty. Policy Soc. 31, 295–306.

Kuzemko, C., 2013. The Energy Security-Climate Nexus: Institutional Change in the UK and Beyond. Palgrave Macmillan, London.

Lundvall, B.-A., 1992. National Systems of Innovation: Towards a Theory of Innovation and Interactive Learning. Pinter, London.

Macedo, I.C., Seabra, J.E.A., Silva, J.E.A.R., 2008. Greenhouse gases emissions in the production and use of ethanol from sugarcane in Brazil: the 2005/averages and a prediction for 2020. Biomass Bioenergy 32, 582–595.

Manzer, L.E., 2013. The role of startup companies in the conversion of biomass to renewable fuels and chemicals. In: Behrens, M., Datye, A.K. (Eds.). Catalysis for the Conversion of Biomass and Its Derivatives. [s.l: s.n.]. p. 43–59.

Milanez, A.Y., et al., 2015. De promessa a realidade: como o etanol celulósico pode revolucionar a indústria da cana-de-açúcar – uma avaliação do potencial competitivo e sugestões de política pública. BNDES Setorial 41, 237–294.

NREL, 2014. National Renewable Energy Laboratory. Biomass Research—National Bioenergy Center. Retrieved from: <http://www.nrel.gov/biomass/national_bioenergy.html> (17.06.14.).

Nyko, D., Garcia, J.L.F., Milanez, A.Y., Dunham, F.B., 2010. A corrida tecnológica pelos biocombustíveis de segunda geração: uma perspectiva comparada. BNDES Setorial 32, 5–48.

Nyko, D., Valente, M.S., Milanez, A.Y., Tanaka, A.K.R., Rodrigues, A.V.P., 2013. A evolução das tecnologias agrícolas do setor sucroenergético: estagnação passageira ou crise estrutural? BNDES Setorial 37, 399–442.

Peters, M., et al., 2012. The impact of technology-push and demand-pull policies on technical change—does the locus of policies matter? Research Policy 41 (8), 1296–1308.

Schnepf, R., Yacobucci, B.D., 2013. Renewable Fuel Standard (RFS): Overview and Issues. Congressional Research Service. Report for Congress.

Slating, T.A., Kesan, J.P., 2011. Making Regulatory Innovation Keep Pace With Technological Innovation. Wisconsin Law Review, Dec.; Illinois Public Law Research Paper No. 10–34; Illinois Program in Law, Behavior and Social Science Paper No. LBSS11-17. Available from: <http://ssrn.com/abstract=1805008>.

Solecki, M., Scodel, A., Epstein, B., 2013. Advanced Biofuel Market Report 2013— Capacity Through 2016. San Francisco. Disponível em: <http://www.e2.org/ext/doc/ E2AdvancedBiofuelMarketReport2013.pdf>.

Sousa, L.C., de Carvalho, E.B., 2013. Biocombustíveis no Brasil: mudanças institucionais, competitividade e governança federal. Revista Ibero-Americana de Ciências Ambientais 4 (1), 69. 14 ago.

Souza, G.M., Victoria, R.L., Joly, C.A., Verdade, L.M. (Eds.), 2015. Bioenergy & Sustainability: Bridging the Gaps. SCOPE-FAPESP-BIOEN-BIOTA-FAPESP CLIMATE CHANGE. Retrieved from: <http://bioenfapesp.org/scopebioenergy/index.php/chapters>.

U.S. DOE, 2014a. U.S. Department of Energy. About the Bioenergy Technologies Office. Retrieved from: <http://www.nrel.gov/biomass/national_bioenergy.html> (17.06.14.).

U.S. DOE, 2014b. U.S. Department of Energy. National Laboratories. Retrieved from: <http:// energy.gov/maps/doe-national-laboratories> (16.06.14.).

Weiss, C., Bonvillian, W.B., 2009. Structuring an Energy Technology Revolution. The MIT Press, Cambridge, MA.

Wyman, C.E., 2001. Twenty years of trials, tribulations, and research progress in bioethanol technology: selected key events along the way. Appl. Biochem. Biotechnol. 91–93, 5–21. Retrieved from: <http://www.ncbi.nlm.nih.gov/pubmed/11963878>.

Ye, F., Paulson, N., Khanna, M., 2014. Technology uncertainty and learning by doing in the cellulosic biofuel investment. In: Agricultural & Applied Economics Association's 2014 AAEA Annual Meeting. Anais. Minneapolis, MN.

Chapter 5

Innovation in the Brazilian Bioethanol Sector: Questioning Leadership

S. Salles-Filho[1], A. Bin[2], P.F.D. Castro[1], A.F.P. Ferro[3] and S. Corder[1]

[1]Department of Science and Technology Policy, Institute of Geosciences, University of Campinas (UNICAMP), Campinas, São Paulo, Brazil [2]School of Applied Sciences, University of Campinas (UNICAMP), Campinas, São Paulo, Brazil [3]Laboratory for Studies on the Organization of Research & Innovation, University of Campinas, Campinas, São Paulo, Brazil

INTRODUCTION

There is no doubt that the favorable context for biofuels, which have expanded since the early 2000s, opened great perspectives for countries that already had their production and technology consolidated. At first, this would seem to be a great opportunity for a less developed country to become the protagonist in the critical area of energy production and consumption.

However, the matter is not that simple. The technological development trajectory of ethanol in Brazil, whether in the agricultural or industrial area, was developed based on a nonproprietary way, with available codified knowledge, making business processes, practices and models relatively simple to transfer. Additionally, with all these new technological and market opportunities, ethanol has attracted investments in other countries that have less tradition in the sector, but with well-developed technological, productive and market capabilities. However, when a successful history with only local reach, as in the case of ethanol in Brazil, is confronted with a global expansion of interests, a competitive process of the same proportions unfolds.

For example, from 2006 up to now, the USA has surpassed the Brazilian production of ethanol and, since 2012, it has produced twice as much as Brazil (Larrea, 2013). This period also saw an increase in investments in second-generation ethanol (2G ethanol) which is produced from cellulose. Despite these important efforts of Brazil to advance in this new technological path, the USA and other countries have presented remarkable advances. In 2014, there

Global Bioethanol.

were three production mills operating on an industrial scale and two were under construction in the USA, while in Brazil, two were operating on an industrial scale and three others were being built.

Brazil is in a typical race on the technological trajectory concept, as proposed by Giovanni Dosi (1984), where several technologies (and capitals) compete for dominance. Even at the time of writing (2015), it is still unclear which trajectory shall prevail. It is not even clear whether the 2G ethanol will be an alternative or complementary trajectory to the first-generation ethanol, or if it will be an economic reality in the biofuel domain.

The sugarcane and ethanol agro-industrial sectors, with their century-old history in Brazil, have been going through structural changes in the past 15 years. It is currently called the sugar-energy sector, precisely because it comprises ethanol production and electricity generation—this last by means of burning bagasse. Energy is, in many cases nowadays, the main business of this sector (Poppe and Cortez, 2012).

The fact that the sector has started to emphasize the "energy" term instead of "ethanol" reveals a moment of sectorial inflexion. This is followed by the change in the structure of capitals in the sector and a new technological and compositional perspective in the energy matrix. As is known, ethanol fuel is currently very important in two countries: the USA and Brazil. The former uses it mixed with gasoline and the latter, as well as adding it to gasoline, uses it directly in the fuel tanks of vehicles and, more recently, of jet aircrafts.

Brazil has had a leadership trajectory in the production and use of biofuels that has distinguished it globally from other countries—at least until the beginning of the 2000s—whether in the genetic breeding of sugarcane crops or in the industrial processes of fermentation and distilling. In the past 10 years, more countries and companies have started investing in technologies to make ethanol feasible from sources other than sugarcane, especially from cellulose, placing the 2G ethanol theme on the global agenda.

This is a sector with a market structure concentrated and organized in large economic groups (both local and multinational), intensive in scale and usually adopting innovations from other sectors, especially from upstream industries. Historically it has always behaved as a supplier-dominated sector (according to Pavitt's typology; Pavitt, 1984): it acquires or adopts technologies developed by other companies, be it from public or private research centers.

In a certain way, it is also a scale-intensive sector, since there are important scale factors for both sugar and ethanol due to their production volumes. Therefore, innovations that allow scale earnings have always been present in the innovation rationale of the sector, even though they do not explain the innovation sectorial logic as a supplier-dominated category does.

In this context, the cost factor and the capacity of offering products at a lower cost tend to overcome the other innovation opportunities with potentially greater impact, which have a secondary role in—or are absent from—corporate strategies. As a result, the structures for innovation management also play a secondary role when compared to the support and continuous (and incremental) improvement structures on productive processes.

This situation presents a few hindrances for the promotion of innovation and innovative management roles in this sector, as it demands the displacement of more conservative behaviors.

Recently, with the advent of 2G ethanol and the entrance of large energy, chemical, oil and gas and petrochemical groups in Brazil, the sector is going through a certain behavioral change, and it is starting R&D and introducing innovations developed by the companies themselves. The typical passive behavior of a supplier-dominated market has started to be punctuated with more active behaviors, ranging from an interactive relationship with specialized suppliers to the development of their own R&D (science-based type).

The main objective of this chapter is to discuss the actual situation of practice and trends of technological and nontechnological innovation in the bioethanol sector in Brazil framed by the international scene of biofuel perspectives. Because of these topics, the manuscript is based on a field study carried out from 2013 to 2014 involving 35 companies that represent about one third of all Brazilian bioethanol production. This study gathered data and opinions about the extent to which these companies are dealing with the innovation challenge and how new areas of knowledge and new technologies have being considered in their strategies.

Therefore, the chapter is organized into four parts: prior literature on the futures of bioethanol (and of biofuels); methodology; results and main findings; and final remarks.

MAIN TRENDS

In the past 10 years considerable attention has been given to biofuels and their perspectives as sources of energy (Gardner and Tyne, 2007; Trindade, 2009; Moschini et al., 2012; Larrea, 2013). Global biofuel production has increased approximately sixfold between 2000 and 2010. Bioethanol and biodiesel are the most important products amongst biofuels. However, bioethanol alone represents more than 70% of total biofuel production and consumption in the world, mainly because of recent policies in the USA (Rajagopal et al., 2010) and in Brazil (Leite et al., 2009; Hira and Oliveira, 2009; Furtado et al., 2011). Biodiesel has increased in Europe but it is still far from the levels of bioethanol production (Moschini et al., 2012).

Whilst the USA has increased its bioethanol production between 2000 and 2012 from around 6 billion liters to more than 50 billion liters, in Brazil, in the same period, production has increased by about half of that. In 2006, the USA surpassed the Brazilian production and by 2012 it was twofold greater (Larrea, 2013; Salles-Filho, 2015). These countries represent the two biggest producers worldwide but they have built very different paths both in terms of policies and of technologies (Moschini et al., 2012).

While the US reached that level of production as a result of an internal regulatory mandate intending to decrease the emission of greenhouse gases and other contaminants of fossil fuels, Brazil increased its production based on a wider array of different drivers (Leite et al., 2009; Furtado et al., 2011; Poppe and Cortez, 2012). In terms of technologies, the USA is the greatest producer of bioethanol made from cornstarch while Brazil is the major bioethanol producer based on sugarcane. Around 11% of the corn grown in the USA is used as raw material for bioethanol production.

In both countries bioethanol is mainly consumed by the internal markets and there is a low level of international trade of bioethanol nowadays (around 10% of global production). That suggests an interesting situation: bioethanol cannot yet be considered a global commodity as it is strongly linked to national decisions and regulations.

Taking a US case, the Renewable Fuel Standard—a program managed by the Environmental Protection Agency—states that by 2022 the US consumption of renewable fuels should achieve 25% of total fuel consumption. That means a humongous volume of 120 billion liters per year, or more than twice the present level (Milanez et al., 2015). In this case, the internal regulation is a powerful driver and the internal market the main target.

In the Brazilian case there was not such a strong biofuel consumption enforcement. However, there were internal decisions of stimulating the production and consumption of an alternative nonfossil fuel, particularly for two main and complementary reasons. The first refers to the historical importance of the sugar-ethanol sector in the country, which allowed initiatives of employing bioethanol as a substitute for gasoline. This started to take place in the the 1930s, but was definitely stimulated as a national policy in the middle of the 1970s. More recently, with the introduction of flex fuel engines, ethanol became a major source of energy and represents circa 40% of liquid fuels consumed by cars. Secondly, and precisely because of the previous reason, Brazil has developed a huge number of scientific and technological capabilities throughout public and private research organizations (Furtado et al., 2011; Campos et al., 2015).

This situation has entailed high expectations about the role Brazil could play in biofuels. At the onset of the 2000s, the country had the biggest sugarcane-based bioethanol industry in the world, alongside well-established and recognized research organizations. As in a few moments in history a less developed country emerged as a potential global player because of its capabilities, accumulated in a critical area such as energy production (Leite et al., 2009; Crago et al., 2010; Furtado et al., 2011). This was an opportunity and also a threat.

The question that immediately arises is: would this country be able to take advantage of the biofuel wave and seize global leadership based upon its competencies and unique experience in the ethanol sector?

Many authors have highlighted the Brazilian advantages, stressing its condition as a first-comer (Farrell et al., 2006; Hira and Oliveira, 2009; Furtado et al., 2011; Macrelli et al., 2012). Others have pointed out the challenges of environmental and social sustainability as a precondition to develop the bioethanol sector in a global perspective—and then to take leadership (Cortez et al., 2002; Leite et al., 2009; Martinelli and Filoso, 2008; Trindade, 2009; Hall et al., 2011).

Very few have discussed the extent of those advantages, based on experience and even on basic and applied research, and if they were strong enough to support a sustainable development and to keep global leadership. João et al. (2012) and Salles-Filho (2015) have shown the weaknesses of the Brazilian bioethanol innovation system.

These authors stress two main reasons for that: nonproprietary and codified knowledge and unclear and conflictive policies. João et al. (2012) carried out a study in which they showed that Brazilian scientific and technological production measured by scientific articles published in indexed journals and by patents filed in different patent offices worldwide, present figures that are far from those of the USA, Canada, France, and Italy. This study led to a comparative analysis of the scientific and technological production in cellulosic bioethanol, widely considered the most important frontier of ethanol production (Cardona et al., 2010; Dias et al., 2011; Macrelli et al., 2012; Larrea, 2013; Milanez et al., 2015). In this regard, Dal Poz and Silveira (2015) showed a similar scenario: the scientific and technological achievements of cellulosic bioethanol have been led by European and North American companies and research organizations. In this play, Brazil has been, at most, a supporting actor. Even in the first generation, as we are going to show in this manuscript, there are concerns about the internal capacity to face the challenges of assuming the leadership in bioethanol.

In the following topics we will present the contents and the main findings of a field study about how companies are dealing with the challenge of becoming innovative in this sector. Data presented herein are the result of interaction with companies participating in the NAGISE Program[1] throughout 2013 and 2014.

Data and information are related to a universe of 35 industrial mills representing 58 different companies (national and multinational) that represent about one-third of Brazil's ethanol production.[2]

STUDYING INNOVATION IN BIOETHANOL

The NAGISE Program was made up of three main activities: (1) training on innovation management that involved approximately 80 graduate professionals from sugar-energy sector companies; (2) an innovation survey applied to participating companies; and (3) technical support for the companies to elaborate their strategic plans on innovation and innovation management. In this sense, NAGISE is a pioneer, systematically and methodologically structured initiative

---

1. NAGISE is the acronym of Laboratory of Innovation Management in the Sugar-Energy Sector. It is the result of a Call for Proposals opened by the Brazilian Innovation Agency, whose objective was to structure Innovation Management Centers in Brazil. NAGISE was created in 2011 by a network of partners that encompassed teams from the University of Campinas—UNICAMP (main institution), Brazilian Agricultural Research Corporation—EMBRAPA, Federal University of Pernambuco—UFPE, Sugarcane Industry Union—UNICA; Agronomic Institute of Campinas—IAC and the University of São Paulo—USP.
2. Data and findings are mostly based upon Bin et al. (2015).

for the development of innovation and innovation management in the Brazilian sugar-energy sector.

Innovation Survey

The methodology employed for data collection consisted of the application and analysis of results from a survey directed at NAGISE companies. The survey had three main parts: (1) profile of participating companies; (2) innovation status and efforts; and (3) innovation management. The survey was mainly based on the Brazilian Innovation Survey (Pintec),[3] which, in its turn, is based on the Community Innovation Survey (CIS) and Oslo Manual (OECD, 2005; Jankowski, 2013).

Thus, the survey presented structured questions on R&D efforts, innovation, cooperation, partnerships, technological and nontechnological outputs and outcomes. It was answered by the NAGISE's participants on a web-based platform especially customized for this purpose.[4]

In order to develop Part 3 of the questionnaire, on innovation management, which was absent in the Pintec research, inspiration was found in works discussing the managerial competences of innovation and creation of innovation/ innovative culture within the corporate scope, such as Tidd et al. (2008), Phaal et al. (2006), Dodgson et al. (2008) and Hidalgo and Albors (2008); and in works directed more towards the measurement of innovation management, such as Adams et al. (2006). The innovation/innovative management indexes were also placed in a template filled out by the companies on the same web platform.

Innovation Challenges

The NAGISE program also identified what the innovation priorities were from the companies' points of view.

The challenges prioritized by the companies were identified in two stages. The first consisted of a survey of bottlenecks and relevant opportunities for the sector as described in the technical and economic literature related to the sector. These bottlenecks and opportunities were grouped into three main blocks that represented the ethanol productive chain: agriculture; industry; and distribution/ marketing. For each block priorities were depicted in three domains of innovation: technological; marketing and organizational; and institutional.[5]

---

3. A research done every 3 years since 2000 by the Brazilian Institute of Geography and Statistics (IBGE), with support from the Brazilian Innovation Agency and the Ministry of Science, Technology & Innovation (MCTI).
4. It is important to highlight that innovation management experts (tutors) monitored the data collected on the web platform. All data presented by the companies were checked by experts, discussed and validated directly with the companies' participants.
5. On technological, marketing and organizational innovation, the concepts from the Oslo Manual (OECD, 2005) are applied. On corporate innovation, indexes such as creation of laws, regulatory and political milestones and other forms of institutional milestone changes were considered.

Before participants went to rank the initial list of challenges, the list was validated by 50 experts by means of a web consultation. Based on the suggestions received, a definitive list was established. This list was then forwarded to the participants in the NAGISE Program, asking them to prioritize and complete the list, as necessary.

The prioritization made by the participants took into consideration the degree of impact expected from each challenge in the following criteria: (1) increase of the physical productivity of the operations in the sector; (2) reduction of operational costs; (3) generation of new business models; (4) generation of social-environmental benefits; (5) technical feasibility; (6) period of maturing; and (7) aggregated importance for the sector (the sectorial perspective).

This prioritization was made in the same platform used for the survey. It was a qualitative prioritization based on a 5-point semantic scale ranging from 0 (null importance for the proposed challenge) to 4 (maximum importance for the proposed challenge).

WHAT IS GOING ON IN THE BRAZILIAN CASE

The results from the innovation survey and the priority challenges are presented below. This is an unprecedented analysis on what is really being done in terms of innovation in the bioethanol sector in Brazil. Before presenting the results, it is important to show the profile of the companies participating in the NAGISE program.

Profile of the Companies Surveyed

Data presented for the analyses of innovation and innovation management of companies in the sugar-energy sector were based upon the answers from 35 productive units from different segments in the sugar-energy chain, representing 27 different companies.[6] The following profile shows characteristics of these 27 companies.

In that sample of companies there were 23 out of 27 producing sugar, ethanol and byproducts (bagasse, yeast, bioelectricity), two companies dedicated to biotechnology and two others dedicated to specialized technical services (engineering consultancy). Three companies were located in the northeast, two in the central-west and 22 in the southeast region, with a strong prevalence in the state of São Paulo.

This is a portrait of the distribution of this sector in Brazilian territory, which has a predominance of sugar and bioethanol production in the southeast region. This region is responsible for 91% of the sugar production and 93% of the ethanol production in the country (UNICA, 2013). The state of São Paulo alone has

---

6. Actually, the NAGISE program worked with 80 professionals, from 35 industrial mills, that worked for 27 companies that represented 58 industrial groups during 2013 and 2014.

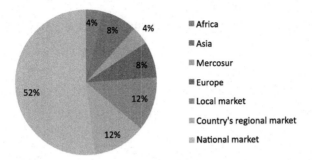

FIGURE 5.1   Bioethanol market destination according to companies.

a concentration of 63% of the sugar production and 51% of the ethanol production in Brazil (UNICA, 2013).

The state of São Paulo is also responsible for the centralization of economic and managerial decisions of the main groups in the sector, which have expanded their productive units mainly towards the central-west, or new groups and foreign companies that have started in the sector, especially since the beginning of the 2000s.

Slightly more than half (14) are independent companies and 13 are a part of groups whose headquarters are located in Europe (3), USA (1) and, mainly, in Brazil (9). A quarter of these participating companies have foreign capital. Companies with foreign capital are controlled by groups from Asia, Europe and the USA. Only three groups/companies are joint ventures between Brazilian and foreign companies/groups.

In terms of revenue (total sales), the main products between 2011 and 2012 were sugar (62%) followed by ethanol. Only 18% of the groups/companies declared ethanol as the first and sugar as the second most important products (only one of the companies is 100% dedicated to the production of ethanol). A third product is quickly gaining importance in the sector—the electricity produced from the burning of bagasse and sugarcane straw. This product is present in 80% of the cases studied, and in a few of them it already represents the second largest source of income, surpassing bioethanol.

The internal market is the main destination of bioethanol (Fig. 5.1): 76% of the groups/companies declared the internal market as the main destination. For companies aiming at external markets, Europe and Asia (8% each), and Mercosur and Africa (4% each) are the most relevant.

The average productivity levels for sugarcane in Brazil are relatively low (less than 70 t/ha, according to CONAB, 2014) when compared to the higher levels seen at some companies in the state of São Paulo (of 100 t/ha). Among the participating companies, 71% have declared an average agricultural productivity higher than the Brazilian average, with the predominance of productivity between 70 and 79 t/ha. Despite this superior productivity among the companies participating in NAGISE in relation to the Brazilian average, it can be said that

TABLE 5.1 Companies' Size According to the Number of Employees

Number of Employees	Number of Companies
1–100	3
101–1000	2
1001–5000	10
5001–10,000	5
10,001–15,000	3
+15,000	2
Total	25

the productivity still falls short of the levels considered satisfactory and feasible, which is around 100 t/ha.

Regarding ethanol productivity, 52% of the groups and companies have declared a productivity of approximately 4 m^3 ethanol/ha, which is below the index of 6 m^3/ha found for the 2011/12 agricultural year in São Paulo (CONAB, 2014).

As expected, most of the groups and companies are large corporations (Table 5.1). The majority have between 1001 and 5000 employees. Five companies/groups have over 10,000 employees. The biotechnology and service companies present a lower level of personnel when compared to the sugarcane and ethanol production plants (up to 320 employees).

On average, for every 23 employees, there is one with higher education level and among the participating companies, there are approximately 10 people with a Masters degree and 4 with PhDs for every 10,000 employees. However, these results are not surprising, given the nature of activities performed (Fig. 5.2).

Technological Innovation in Corporate Strategy

Technological innovation in the sugar-energy industry takes place in three main domains: agricultural, industrial, and transportation and mechanization areas (Fernandes and Lima, 2012). Among the units analyzed, 80% declared having had some type of innovation in the 2011–2012 periods. This percentage is much higher than that found in the national innovation survey for the average industry in Brazil, which is around 36% (IBGE, 2013). This difference is partially explained by the fact that the companies participating in the NAGISE program are large corporations. When this sample is compared to the average large corporations in the Brazilian industry, the difference decreases: IBGE (2013) shows that 56% of large corporations presented product or process innovation in the same period.

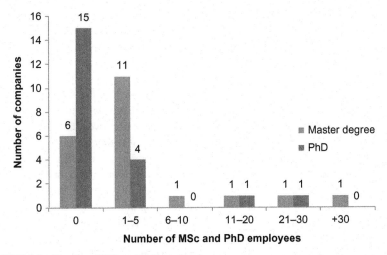

FIGURE 5.2   Number of MSc and PhD employees per company.

From the sample analyzed, only one was exclusively involved in product innovation (less than 3%, compared to 4% on the average of the Brazilian industry) and 10 others in products and processes (approximately 29% against 13.4% in the Brazilian industry). On the other hand, 17 companies have declared innovation exclusively on processes (48.5%, compared to 18.3% in the industry average). These figures show the sugar-energy sector as much more directed to the innovation of processes, although having a similar distribution when compared to the average Brazilian industry.

Coherently, with the sector profile of supplier-dominated and somewhat scale-intensive, most of the innovations introduced, whether they be in process or product, were new to the company itself or to the Brazilian market and were characterized as modernization and a technological update.

Among the innovative products developed by the companies, it is important to emphasize the new sugarcane varieties in the agricultural area. In the industrial sphere, innovations involved the differentiation of sugar types (crystal; very high polarization (VHP); and very very high polarization (VVHP)), the cogeneration of energy from the use of sugarcane straw, and the production of food supplements. There were very few innovation cases of technological innovation in relation to fermentation and distillation processes.

Two "new-to-the-world" innovations were registered: a sugarcane variety for the savanna areas and the implementation of mechanical harvesting methods in adverse topographic conditions. As a rule, the innovations presented were of incremental nature, however, even these are considered new in a global perspective.

Regarding the origin of product innovations, it is interesting to note that most were developed with the participation of the company itself, whether exclusively or with the involvement of other companies or research centers, with only one innovation made abroad. Historically, this is a sector with technological developments made within the country, especially in the agricultural segment of the productive chain.

Process innovations are not only more frequent but also have a more innovative profile when compared to product innovations. On the same line, these innovations are developed mostly by the companies themselves, with little participation by foreign agents. They are mainly related to improvements which originated from production processes, called focusing devices by Rosenberg (1982).

Among the main process innovations, it is important to mention the agricultural area, the improvement of planting methods; soil preparation; introduction of monitoring and production control methods; introduction of agricultural machinery and equipment, mostly destined for harvesting and sowing; and new forms and methods for collecting straw. In the industrial area, it is worth noticing the introduction and automation of machinery and equipment; new types of boilers to increase efficiency in energy generation; and new wastewater treatment systems.

A closer look at the data, having as a background the corporate innovation in the Brazilian context, allows us to make a few complementary observations. The sugar-energy sector reproduces the Brazilian standard according to Pintec (IBGE, 2013), whereas there is a higher predominance of process innovations in relation to product innovations. There is also a similarity in the innovation scope: most companies introduce new products and processes which are new for the company level, with a low innovation rate for the country level (approximately 7% for processes and 20% for products).

The productive plants are dedicated to industrial activities (sugar and alcohol production) and agricultural activities (sugarcane production). They innovate much more as an answer to immediate challenges of cost reduction and environmental constraints, without important incentives to exploit new technological frontiers. Most innovations in the past 5 years came to comply with environmental regulations. This is the case for harvest mechanization, water reuse, waste treatment, cogeneration of electricity and heat to mention the most common. With some exceptions, these were innovations made through the acquisition of technologies already available on the market or developed by suppliers upstream and downstream of the productive chain. There are very few radical and transforming innovation projects in the technological profile of the sector and those are concentrated in approximately three companies.

Generally speaking, one can state that technological and market challenges that are now in development in the biofuel segments around the world are not in the strategic prospect of this sector in Brazil.

Most of the priorities ranked by the companies demand very little new knowledge or in-house R&D. In situations in which there is some sort of external pressure for innovation (such as the regulatory mandates) companies seek outsourced services, mainly suppliers.

Regarding R&D efforts, amongst the 35 production units, 18 indicated some in-house activity. Considering employees with higher education degree dedicated to R&D, there are 122 full-time professionals, and from these approximately 20 have a Masters or a PhD, which represents an average of over one postgraduate professional per productive unit. On the other hand, if one takes the whole sample analyzed in the study, the number of researchers with a postgraduate degree would be about 0.6 researcher per productive unit.

In the case of the R&D carried out on second-generation ethanol, only 3 out of 58 groups and companies were directly involved in the technological development of cellulosic ethanol.

It is important to emphasize that the acquisition of machinery and equipment is also the innovative activity considered the most important in the Brazilian industry context, according to data from IBGE (2013). From the point of view of innovative activity importance, the sugar-energy sector presents medium and high relevance for the acquisition of machinery and equipment in 40% of the cases, while on the average for Brazilian industry it is 76%. The importance of the other innovative activities has similar distribution to that found in the Pintec national innovation survey.

A few companies stated that they are part of cooperative arrangements to accomplish innovative activities. However, when questioned on which partnership category tends to be more important, there is an emphasis on universities and research institutes, followed by consultancy companies, corroborating with data found by Spíndola et al. (2012). The professional training centers were also considered to be of medium relevance by nine units.

A view on the figures of intellectual property among the companies of the sample showed a coherent profile. Over a period of 15 years there were 25 patents filed in Brazil and just 3 in the PCT and 26 cultivars registered in Brazil (Table 5.2). Almost all patents and cultivars were filed by a single company (the one involved in technological development to the sector). As for trademarks, one company registered more than 90 brands in Brazil and more than 140 abroad (this is a Brazilian company involved in the diversification of new business models and new markets).

Regarding innovation activities towards the diversification and opening of new markets, most of the companies indicated low to very low relevance. However, 19 out of 35 productive units considered that innovations have had a medium to high impact on supporting the reduction of environmental impact, reinforcing the already mentioned central role of the regulatory framework. The high and medium effects of innovations on reducing production costs was considered by 15 units, and 17 stated there were medium and high impacts on the control of aspects related to health and safety.

This point is further emphasized when compared to the Brazilian scenario: data from Pintec (IBGE, 2013) indicate the innovation impacts on the environment were the least significant for the Brazilian industry during the period

TABLE 5.2 Intellectual Property Applications per Company, in Brazil and Abroad (2000–2013)

Intellectual Property Application	Quantity
Filed cultivar (Brazil)	28
Granted cultivar (Brazil)	28
Filed patents (Brazil)	25
Filed patents (abroad, including Patent Cooperation Treaty (PCT))	3
Filed utility model (Brazil)	1
Filed trademark (Brazil)	95
Filed trademark (abroad)	147
Granted trademark (Brazil)	40
Granted trademark (abroad)	125

between 2009 and 2011, revealing different effort aspects (even though still fragile) toward responsible innovation. The explanation can be found in the regulatory milestone of the sector, which, in Brazil and several other countries, has been pushing for process changes to reduce negative environmental effects of ethanol production and use.

What is really important to notice is the low perception of the commercial gains of innovation. This point only confirms the secondary role of technological innovation in the corporate strategy of companies, especially in the search for productive efficiency.

A large number of units (30) implemented important organizational innovations with emphasis on labor process organization, whether to improve the operational routine and practices or to improve the delegation of responsibilities. The new environmental management techniques were also incorporated by more than half of the units, more precisely, by 57% of them (20 units).

Regarding marketing innovations, nine units considered having made significant changes to the marketing concepts or strategies, and seven also registered significant changes to the design and the esthetics of at least one of the products.

The importance of nontechnological innovations for the sugar-energy sector is not an isolated phenomenon in the Brazilian scenario: IBGE (2013) reveals that innovations of this nature are quite frequent in the Brazilian industry, since over 80% of the companies innovate in products and processes have also introduced organizational marketing innovations, with an emphasis on "new management techniques" and "methods for organizing labor."

Innovation Management

As for the capabilities on innovation management, companies presented a minimalist structure, both in terms of number and specialization of personnel and of organizational features.

The efforts may be considered very low, even in large and highly performance companies. In other words, little of the learning and routines related to innovation management of company headquarters are being transferred to their sugar-energy subsidiaries, even in multinational corporations whose experience in innovation management is widely recognized.

The most proactive units in terms of innovation management are those acting at the knowledge frontier of the sector, that is, in the development of new biomass sources, and second generation. Nevertheless, even in these cases, there is still very little to assess, since the operational stage is embryonic. With few exceptions, such as one of the companies that was founded to develop 2G ethanol processes, it can be stated that the sector presents a very initial effort in developing capabilities in innovation management.

For instance, only 3 out of 35 units had their own full-time staff dealing with innovation management activities, working with small teams of one to five people. In any case, albeit small, these areas concentrate on people with some kind of training in the innovation management subject.

The number of units with people who were partially dedicated to innovation management activities was higher, totaling 10 out of 35 units. In almost all of these units, the innovation/innovative management activities were decentralized in several areas/departments. Only three of them had an innovation management plan able to guide the innovative activities. Twenty-one units stated they did not have any personnel (of their own or outsourced) dedicated to innovation management.

Innovation Challenges According to Companies

Roughly speaking, challenges of the agricultural segment—which represents 70% of the ethanol total production costs—were considered the most relevant by the companies participating in the NAGISE program. Although the set of technological challenges presented a greater weight in preferences, the institutional challenges, as well as the marketing and organizational ones, were well ranked by participants.

The companies understood that one of the main technological bottlenecks for the sector in the agricultural segment was the development of new varieties aiming mainly at productivity and adaptation to nontraditional regions. As shown by Campos et al. (2015) and Furtado et al. (2011), Brazil is a country with large programs in sugarcane R&D, and a large supply of new varieties. The fact that the companies' concerns are headed to sugarcane productivity in a country where new varieties are abundant brings up the question of what is going on with the supply and demand of new technologies. The most probable answer lies in the low levels of investment in the sector and on the conservative behavior towards innovation, particularly on the agriculture side of the chain.

Another challenge related to the same subject is the need to have better quality and healthy seedlings to meet the needs of commercial culture; machinery, implements and chemical inputs for sowing and harvesting sugarcane more appropriate to the new conditions of mechanized harvesting; biological and pest control alternatives with integrated pest control practices; and productivity increase with the development of energetic cultures that can be grown alongside sugarcane.

With some exceptions, most of the companies in the sector reveal chronic problems related to productivity and profitability connected to the existence of the sugar variant[7] and the low growth of ethanol in both the internal and the external markets.

In a medium to high prioritization level, there are challenges such as the use of transgenic and genetic improvement of biological control; development of machinery adapted for small producers; adding value to all sugarcane residues, among others. As it can easily be confirmed in specialized literature, most of these technological challenges are already being sorted out and respective solutions are mostly available or need some investment to be developed at the industrial level.

The main items discussed by the companies in marketing and organizational innovations are related to the dissemination and transfer of techniques and technologies, many of them already existing.

In relation to the industrial segment, companies have prioritized the need to improve the performance of agricultural and industrial logistic systems, especially those related to the systems of cutting, loading and transporting sugarcane. Companies understand the bottlenecks and opportunities in the marketing and organizational area as very relevant and they know these could easily be adopted. The question of why they have not yet done so still remains. Again, the evidences of a slight or small need from the microeconomic point of view, seem to be a plausible explanation: firms do not innovate because they do not need to, given the competitive and market conditions.

Regarding the challenges for institutional innovation, they were considered to be of high relevance, particularly for public policies and regulatory milestones towards sectorial competitiveness.

In sum, from the companies' points of view, priorities are mostly concerned with innovations in fields of knowledge and in activities already well established. Whether in technological or nontechnological bottlenecks and opportunities, firms are particularly interested in modernization and upgrading their

---

7. The sugar variant means that a company in the sector usually has the alternative of producing sugar instead of ethanol, if the relative prices justify the change. This partially explains the conservative behavior regarding the investment in innovation in the sector.

productive structures. There is no such movement towards new technologies, not even explicit concerns to attend to an expanding global market.

FINAL REMARKS

In general terms, it can be concluded from the above discussion that the role of innovation and innovation management within the companies of the Brazilian sugar-energy sector is still limited. Nevertheless, it is important to note that the country has research competences in all technological challenges mentioned herein. There is a sort of "dual" innovation system in the country composed by a well-developed research subsystem and a noninterested productive subsystem.

A survey of the Research Group Directory in the Brazilian National Scientific and Technological Development Council (http://lattes.cnpq.br/web/dgp) reveals the presence of over 140 research groups in Brazilian universities, acting on R&D in all the segments of ethanol productive chain. The country also has at least four large sugarcane breeding programs, two in the public and two in the private sector. In the field of industrial technology research, there are at least three important centers in the country.

Moreover, since the mid-2000s, the country has R&D and innovation financing sources in the three main public agencies supporting research and innovation programs. These are the programs *"Plano de Apoio à Inovação dos Setores Sucroenergético e Sucroquímico"* (PAISS), executed by both the Brazilian Development Bank (BNDES) and the Brazilian Innovation Agency (FINEP), while the Bioenergy program (BIOEN) is financed by the São Paulo Research Foundation (FAPESP). These two programs together may reach investments of approximately US$ 1.8 billion in 5 years.

In this manner, the innovation sectorial system in Brazil is presented in a dual perspective. On one hand, public and private research centers are producing knowledge and technology; and on the other hand, users (sugar and ethanol companies) with a slight relationship with the research system and low adoption levels of technologies and negligible competence on the technology and innovation management subjects.[8]

According to Fernandes and Lima (2012), factors such as low-cost land and labor, as well as a routinized production, help to explain the low technological dynamism of this industry throughout history. Moreover, it is possible to add to these arguments related to costs of capital and the risks of an unstable market, which plays an alternatively between the two main products—sugar and ethanol. Other macroeconomic constraints, such as interest rates and unstable policies, have also contributed to frame this situation.

---

8. For more details on the innovation system in the sugar-energy sector in Brazil, see Campos et al. (2015) and Furtado et al. (2011).

It is not surprising that resources dedicated to innovation and their results are limited, even though part of the companies consider that outsourced research efforts are elevated and somehow connected to their own efforts.

The trajectory being built in this sector is focused on incremental process improvements, without much space for product innovations or even for updates requiring capital renewal. Everything else results from this: there is very little investment in R&D; on the other hand, the relationship with suppliers for the acquisition of machinery, equipment and software is passive, characterizing a supplier-dominated sector.

Although the relationship with research institutes and universities is frequent, this is not translated in economic outputs and outcomes. This perception is reinforced by the low level of protection of intellectual property rights and the negligible effort to get external financing sources for innovation (including governmental support).

Finally, the innovation management structures, if and when they exist, are minimalist. It can be said that the sector is attempting to change, but it does not yet prioritize the search for innovation or a greater interaction with research institutes in order to complement its slight or nonexistent investment in their own R&D departments (Fernandes and Lima, 2012). This is also not surprising when one analyzes the general behavior of Brazilian industry in terms of innovation. As mentioned throughout the text, much of what has been observed in the sugar-energy sector is also common to other sectors.

The sugar-energy sector is one of the main segments collaborating with the prominent position of the Brazilian agribusiness. Nonetheless, several challenges associated with the different segments in the chain still need to be overcome in order to resume new productivity rates and competitiveness gains.

The overcoming of such challenges invariably goes through both technological and nontechnological innovations. Only the companies that have invested (and still invest) in innovation throughout the chain will be able to face the challenges raised by the second generation and the possible expansion of the international market for bioethanol.

For the agricultural segment, it was observed that most of the challenges are related to well-known technological issues, as well as to fields of broad competence in Brazil. Most of these challenges could be considered as "old" ones. Actually, they do represent chronic problems related to low investment in modernization and in innovation.

It can be observed that the current investment levels in innovation in the sector are low for any type of demand. Several companies in the sector outsource R&D to research companies and organizations, but even those are not frequent adopters.

Although investments have expanded in the country particularly because of the above-mentioned programs PAISS and Bioen, and despite the presence of companies that are really engaged in technological and nontechnological advances in both the first and second generations, the sectorial analysis shows figures that are inconsistent with the demands of a changing scenario such as the one currently being experienced.

Brazil is not prepared for the success of second-generation bioethanol. Most likely, an eventual success of the ongoing second-generation initiatives may either reinforce or strongly disrupt the sector. Despite the possibilities of living together, 1G and 2G ethanol may also compete with each other. That will depend on the relative prices, which can only be more or less speculated on for the time being.

In the past few years the sugar-energy sector has experienced a sort of vicious cycle: low investment in modernization and innovation results in low economic performance that results in still lower investments and so on.

The innovation management initiatives within the companies in the sector are embryonic, albeit important. Nonetheless, the challenges presented to the sector cannot waive a strengthening of innovation competences. The country currently offers affordable and appropriate financing for innovation and the companies are not taking advantage of this.

The argument for not investing in innovation because it is expensive and inappropriate in a time of crisis such as the one the sector has been going through in recent years is what sustains the vicious cycle. First, because in the good times there was no expansion in innovation investments; secondly because it is exactly through innovation that one can leave the limited productivity and cost thresholds to create more favorable competitiveness conditions.

To halt this vicious cycle, more effort on bringing universities, research centers and companies together may trigger a more innovative environment based on applied knowledge. This approach may provide the necessary support to ethanol 2G development. Besides, strategies headed to the external market are absolutely necessary in order to get essential stimulus to innovate that only the global market can effectively assure in this area.

REFERENCES

Adams, R., Bessant, J., Phelps, R., 2006. Innovation management measurement: a review. Int. J. Manag. Rev. 8 (1), 21–47.

Bin, A., Ferro, A.F., Castro, P.D., Corder, S., Lemos, P., 2015. Diagnóstico e Prognóstico da Inovação e da Gestão da Inovação: onde estamos e para onde vamos? In: Salles-Filho, S. (Coord.) Futuros do Bioetanol: o Brasil na liderança? Elsevier, Rio de Janeiro, pp. 145–171.

Campos, A., Lucafó, B., Corder, S., 2015. Sistema de Inovação do Setor Sucroenergético no Brasil. In: Salles-filho, S. (Ed.), Futuros do Bioetanol: o Brasil na liderança? Elsevier, Rio de Janeiro, pp. 35–54.

Cardona, C.A., Quintero, J.A., Paz, I.C., 2010. Production of bioethanol from sugarcane bagasse: status and perspectives. Bioresour. Technol. 110 (13), 4754–4766.

CONAB – Companhia Nacional de Abastecimento, 2014. Cana-de-açúcar – Brasil: Série Histórica de Produção de Cana-de-Açúcar: Safras 2005/06 a 2014/15 em kg/ha. Available at http://www. conab.gov.br/conteudos.php?a=1252&t=&Pagina_objcmsconteudos=2#A_objcmsconteudos (accessed 16.06.14.).

Cortez, L.A.B., Griffin, M., Scandiffio, M.I.G., Scaramucci, J.A., 2002. Worldwide use of ethanol: a contribution for economic and environmental sustainability. In: Sustainable Development of Energy, Water and Environment Systems Proceedings, Dubrovnik.

Crago, C.L., Khanna, M., Barton, J., Giuliani, E., Amaral, W., 2010. Competitiveness of Brazilian sugarcane ethanol compared to US corn ethanol. Energ. Policy 38, 7404–7415.

Dal Poz, M.E., Silveira, J.M.F.J., 2015. Trajetórias Tecnológicas do Bioetanol de Segunda Geração. In: Salles-Filho, S. (Ed.), Futuros do Bioetanol: o Brasil na liderança? Elsevier, Rio de Janeiro, pp. 111–124.

Dias, M.O.S., Cunha, M.P., Jesus, C.D.F., Rocha, G.J.M., Pradella, J.G.C., Rossell, C.E.V., et al., 2011. Second generation ethanol in Brazil: can it compete with electricity production? Bioresour. Technol. 102 (19), 8964–8971.

Dodgson, M., Gann, D., Salter, A., 2008. The Management of Technological Innovation: Strategy and Practice. Oxford University Press, New York.

Dosi, G., 1984. Technical Change and Industrial Transformation: The Theory and an Application to the Semiconductor Industry. MacMillan, Londres.

Farrell, A.E., Plevin, R.J., Turner, B.T., Jones, A.D., O'Hare, M., Kammen, D.M., 2006. Ethanol can contribute to energy and environmental goals. Science 311, 506–508.

Fernandes, A.C., Lima, J.P.R., 2012. Os labirintos da interação Universidade-Empresa: apontamentos a partir de dois estudos de caso (elétrico e sucroalcooleiro) em Pernambuco. Estudos Universitários (UFPE) 31 (12), 73–92.

Furtado, A.T., Scandiffio, I.G.S., Cortez, L.A.B., 2011. The Brazilian sugarcane innovation system. Energ. Policy 39, 156–166.

Gardner, B., Tyne, W., 2007. Explorations in biofuels economics, policy, and history: introduction to the special issue. J. Agr. Food Ind. Organ. 5 (2), 1–6.

Hall, J., Matos, S., Silvestre, B., Martin, M., 2011. Managing technological and social uncertainties of innovation: the evolution of Brazilian energy and agriculture. Technol. Forecast. Soc. 78, 1147–1157.

Hidalgo, A., Albors, J., 2008. Innovation management techniques and tools: a review from theory and practice. R&D Management 38 (2), 113–127.

Hira, A., Oliveira, L.G., 2009. No substitute for oil? How Brazil developed its ethanol industry. Energ. Policy 37, 2450–2456.

IBGE – Instituto Brasileiro de Geografia e Estatística, 2013. Pesquisa de Inovação – PINTEC 2011. IBGE, Rio de Janeiro.

Jankowski, J., 2013. Measuring innovation with official statistics. In: Link, A.N., Vonortas, N.S. (Eds.), Handbook on the Theory and Practice of Program Evaluation. Edward Elgar, Cheltenham.

João, I.S., Porto, G.S., Galina, S.V.R., 2012. A posição do Brasil na corrida pelo etanol celulósico: mensuração por indicadores C&T e programas de P&D. Revista Brasileira de Inovação 11 (1), 105–136.

Larrea, S., 2013. Fiscal and Economic Incentives for Sustainable Biofuels Development: Experiences in Brazil, the United States and the European Union. Inter-American Development Bank, RG-K1128.

Leite, R.C.C., Leal, M.R.L.V., Cortez, L.A.B., Griffin, M., Scandiffio, M.I.G., 2009. Can Brazil replace 5% of the 2025 world gasoline demand with ethanol? Energy 34, 655–661.

Macrelli, S., Mogensen, J., Zacchi, G., 2012. Techno-economic evaluation of 2nd generation bioethanol production from sugar cane bagasse and leaves integrated with the sugar-based ethanol process. Biotechnol. Biofuel. 5 (22), 1–18.

Martinelli, L.A., Filoso, S., 2008. Expansion of sugarcane ethanol production in brazil: environmental and social challenges. Ecol. Appl. 18 (4), 885–898.

Milanez, A., et al., 2015. De promessa a realidade: como o etanol celulósico pode revolucionar a indústria da cana-de-açúcar – uma avaliação do potencial competitivo e sugestões de política pública. BNDES Setorial 41, 237–294.

Moschini, G.C., Cui, J., Lapan, H., 2012. Economics of biofuels: an overview of policies, impacts and prospects. Biobased Appl. Econ. 1 (3), 269–296.

OECD – Organisation for Economic Co-operation and Development, 2005. Oslo Manual: Guidelines for Collecting and Interpreting Innovation Data, third ed. OECD Publishing, Paris.

Pavitt, K., 1984. Sectoral patterns of technical change: towards a taxonomy and a theory. Res. Policy 13, 343–373.

Phaal, R., Farrukh, C.J.P., Probert, D.R., 2006. Technology management tools: concept, development and application. Technovation 26 (3), 336–344.

Poppe, M.K., Cortez, L.A.B. (Eds.), 2012. Sustainability of sugarcane bioenergy. Center for Strategic Studies and Management (CGEE), Brasília, DF.

Rajagopal, D., Hochman, G., Zilberman, D., 2010. Multicriteria Comparison of Fuel Policies: Renewable-Fuel Mandate, Emission Standard, and Carbon Tax. Institute of Environment, University of California, Los Angeles.

Rosenberg, N., 1982. *Inside the Black Box: Technology and Economics.* Cambridge University Press, Cambridge.

Salles-Filho, S., 2015. Futuros do Bioetanol: o Brasil na liderança? Elsevier, Rio de Janeiro.

Spíndola, F.D., Lima, J.P.R., Fernandes, A.C., 2012. Interação Universidade–Empresas: o caso do setor sucroalcooleiro de Pernambuco. Anais do XVI Encontro Nacional de Economia Política, Rio de Janeiro.

Tidd, J., Bessant, J.R., Pavitt, K., 2008. Innovation Management, third ed. John Wiley & Sons Ltd, West Sussex.

Trindade, S., 2009. The sustainability of biofuels depends on international trade. Energy Sourc. 31, 1680–1686.

UNICA – União da Indústria de Cana-de-Açúcar, 2013. Dados e cotações. Availabe at http://unica.com.br (accessed 05.04.13.).

Chapter 6

Policies Towards Bioethanol and Their Implications: Case Brazil

L.A.B. Cortez[1,2] and R. Baldassin, Jr.[3]

[1]Faculty of Agricultural Engineering, UNICAMP, Campinas, São Paulo, Brazil
[2]International Relations, UNICAMP, Campinas, São Paulo, Brazil [3]Interdisciplinary Center for Energy Planning – NIPE, UNICAMP, Campinas, São Paulo, Brazil

HISTORICAL DEVELOPMENT OF THE BRAZILIAN ETHANOL PROGRAM (PROÁLCOOL)

Brazil has used sugarcane since the 16th century when it was introduced, becoming an important economic vector in the colonial period, particularly in the northeast region. In the initial centuries after colonization, the main primary energy for Brazil was wood.

In general terms it can be stated that Brazil can be considered an "energy-poor" country, at least considering the main global primary energies (coal, oil, gas and more recently shale gas). Brazil has relatively small known coal reserves. Small quantities of oil started to be discovered after the creation of Petrobras in 1954, and more recently relatively important oil reserves were found in deep waters (pre-salt) of the Atlantic Ocean near the coast of Rio de Janeiro and São Paulo.

Due to weakness of the country's energy security, and also to help the sugar sector, ethanol fuel was first introduced in the 1930s by the Vargas government. The IAA (Institute of Sugar and Alcohol) was also created to regulate sugar and ethanol production and, as consequence, mandates were issued to formalize the use of ethanol/gasoline blends. Then, later, the use of ethanol was intensified during the oil crisis of the 1970s resulting in the creation of PROÁLCOOL in Nov. 1975. At that time imported oil responded by 80% of overall oil consumption representing nearly 50% of all imports.

For the Brazilian government, it was obvious to encourage domestic production of liquid fuels to decrease oil dependency, save hard currency, and at the same time create jobs and help to decrease regional socioeconomic inequalities in the country. Although PROÁLCOOL gave substantial positive results,

Global Bioethanol.

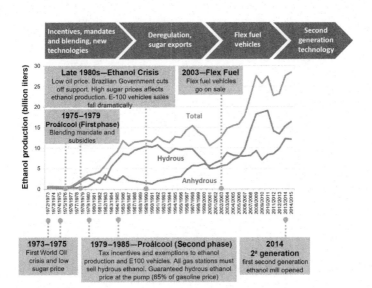

FIGURE 6.1   Different phases of the Brazilian Ethanol Program (1975–2015). *Datagro, 2008. Based on UNICA and ICONE (*Datagro, 2008*), elaborated by ICONE and UNICA.*

both from the gasoline substitution and the socioeconomic impacts points of view, the 40 years that have passed since its creation were marked by ups and downs indicating that also in Brazil the introduction and use of biofuels was not always easy (Fig. 6.1). More about the history of PROALCOOL is available at Copersucar (1989), Rosillo-Calle and Cortez (1998), Goldemberg et al. (2004), FAPESP (2007), BNDES (2008), CGEE (2012), Boeing (2013), Cortez (2014a) Moraes and Zilberman (2014) and Cortez et al. (2016).

Today, Brazil is the second largest producer of bioethanol and the largest producer of bioethanol from sugarcane, which contributes to a relatively clean energy matrix (nearly 50% of energy used in Brazil is renewable). In its energy planning, Brazil recognizes that sugarcane will still play an important role in its energy matrix, both for the production of ethanol and bioelectricity. It is forecasted by MME-EPE (2007) that in 2030 nearly 18.5% of Brazilian primary energy will come from sugarcane products (Fig. 6.2).

Worldwide, the ethanol production scenario seems to be dominated by the United States and Brazil. However, due to the greater availability of land, Brazil has better prospects to expand its production without using the land currently used for food production. There is also good potential in other Latin American countries, such as Colombia and several African countries, but this will depend on how these countries handle their respective political barriers, such as those related to land tenure and the constant threat offered by fossil fuels. In Fig. 6.3 the present world production of ethanol as well as the short- and medium-term prospects can be observed.

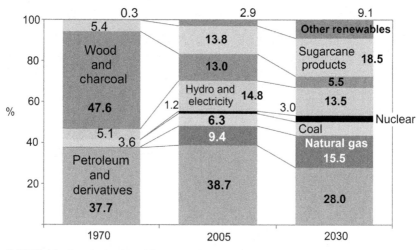

FIGURE 6.2 Prognostic of Brazilian energy matrix: the participation of sugarcane products. *EPE-MME (2010).*

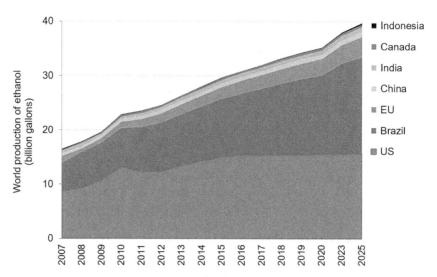

FIGURE 6.3 World production of ethanol between 2007 and 2025. *FAPRI 2008 and 2012, US and World Agricultural Outlook.*

Regarding future markets for biofuels, according to Fulton (2013), significant portions of the main transport sectors will be supplied by biofuels (ethanol and biodiesel mainly). This means by 2050, 25% of light-duty vehicles (250 million tonnes of oil equivalent (Mtoe)), 25% of urban trucks (100 Mtoe), 20% of long-haul trucks (150 Mtoe), 20% of shipping (100 Mtoe) and 20% of aviation (100 Mtoe), totaling around 700 Mtoe (Fig. 6.4).

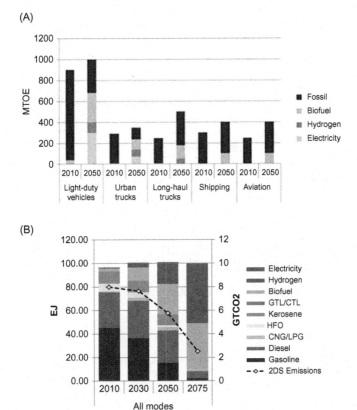

FIGURE 6.4   (A) Estimated participation of fossil, biofuels, hydrogen, and electricity in transport by 2050; (B) Aggregated global transport energy use, IEA 2DS. *Fulton (2013).*

Therefore, it can be stated that there will be a substantial market for biofuels (ethanol, biodiesel and others) in the coming decades. The question now is how they can be produced on a large scale without subsidies? For instance, Brazil learned how to make the first step towards a fossil fuel substitution but "something" is holding back its full expansion. What is it? Is it really only low oil prices or a general criticism about biofuels versus food debate that is the reason for its progress? In the next sections of this chapter we will try to show that it is more challenging than that.

THE CREATION OF THE "BRAZILIAN SUGARCANE ETHANOL MODEL"

Brazil developed its sugarcane ethanol program (PROÁLCOOL) in a very unique way: first using the sugar residue molasses to produce ethanol and later

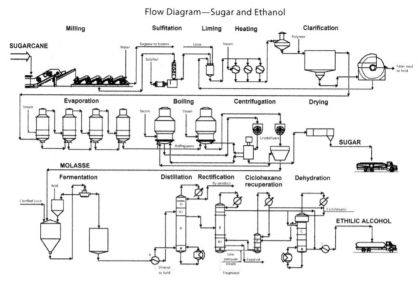

FIGURE 6.5 Flow diagram of the sugar-ethanol and electricity industrial production model used in Brazil. *Finguerut (2005).*

completely integrating sugar and ethanol production to take advantage of its integration. The so-called "Brazilian model" is best described by Finguerut (2005). According to Finguerut, "the Brazilian way for making sugar and alcohol simultaneously has some advantages compared to other sugar countries because less energy is used, the quality of the products is better, and the yields attained are greater (stoichiometric ethanol yield 91%), with lower costs (about US$0.20/liter ethanol)". Technically, at the mill, the integration can be shown using the flow diagram in Fig. 6.5.

Finguerut (2005) also describes in more detail the technical advantages that permit a series of maneuvers that are virtually impossible if the distillery is separated. He also indicates the positive technical advantages that derives from integrating sugar and ethanol production:

1. Lower production costs of sugar and ethanol (due to the high stoichiometric ethanol yield around 91% for the past 7 years, compared with values from 80% to 87% for conventional distilleries);
2. Better quality of both products, especially the sugar quality that can be kept almost constant, even with low-quality sugarcane;
3. Surplus energy that can be sold as electricity to the grid or as fuel bagasse for other industries;
4. Optimization of critical agronomical operations such as harvesting, enhancing industry stabilization and optimization of milling;
5. Profitability even with low sugar prices.

However, the so-called Brazilian model is not only restricted to a mill integration between sugar and ethanol production. It goes beyond that. Actually the Brazilian model is composed of a series of strategies covering the mill itself, the blending and the final consumption, in a way that it gives an integrated inherent "overall flexibility" to the main stakeholders: farmers, mill owners, Petrobras, the Brazilian government and fuel consumers.

The created model presented different levels of flexibility and is described below.

1. The first level (Flex I), at the mill, where it can be choose to produce more sugar or ethanol, depending on the market prices (described by Finguerut, 2005). Today, many Brazilian mills present levels of flexibility for sugar/ethanol production of 40–60%. This means that, depending on market conditions, the mill can switch from a 60% sugar–40% ethanol to a 40% sugar–60% ethanol.

2. The second (Flex II) is composed of two parts. The first still at the mill, in which there is the option to produce hydrous or anhydrous ethanol, depending on the price and opportunities. This has to do with market conditions (price and demand) and is affected by the government regulatory system. The second part of Flex II has to do with government decision to use higher percentages of anhydrous ethanol in the gasoline. Ethanol blends with gasoline can vary. At the beginning of the PROÁLCOOL the practice was E10 (10% ethanol into the gasoline) and then it increased with time. Today the blend is 27% of anhydrous ethanol into the gasoline (E27). Therefore, combining parts one and two, we get "Flex II," representing an advantage for the mill and the government, in this case.

3. The third (Flex III) at the consumer level which can choose to use E27 or E100 at the pump station. The existing engines in Brazil are built to tolerate relatively high ethanol–gasoline blends. The consumers, in practice will have the possibility to use a fuel (E27) that is usually higher in price (today nearly US$1.00/liter) but gives a better mileage (roughly 25% more) to a fuel (E100) that is usually cheaper (at the time of writing nearly US$0.70) but gives a relatively lower performance. Consumers can use E27 or E100 or even mix the two at any proportion.

4. More recently the scheme has also integrated electric energy production and sale, which becomes another important source of income for the mills. The sale of electricity will, more and more, represent another flexibility level, now involving the cane fiber (bagasse and trash) and no longer the cane sucrose. Considering this as another flexibility level the mill would have the option to produce more or less electricity depending, on the one hand, on the raw material fiber content, and on the other hand, on electricity prices and advantages.

This "overall flexibility," combining the mill sugar-ethanol (Flex I), with hydro-anhydrous at the mill combined with anhydrous blend level (Flex II) and flex engine with consumer decision (Flex III) gives a very robust condition for

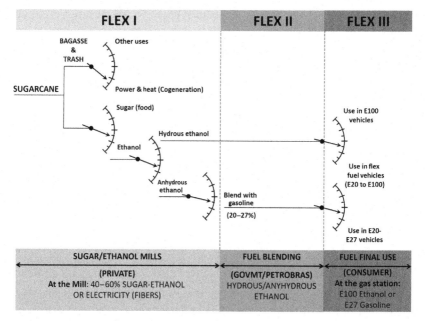

FIGURE 6.6   Schematic of the Brazilian model: production, mixture and final use. *Brito Cruz, C.H., Cortez, L.A.B., Souza, G.M., 2014. Biofuels for transport. In: Future Energy: Improved, Sustainable and Clean Options for Our Planet. Elsevier, London. , 716 pp., ISBN 9780080994246. http://find. lib.uts.edu.au/;jsessionid= B944BE4C8E2E60E5C1BDB7274F25240F?R= OPAC_b2845998.*

the stakeholders besides significantly improving the situation for the country, combining policies in the sugar sector with the fuel sector. Fig. 6.6 gives an illustration of these three flexibilities.

However, the sugarcane ethanol industry has developed successfully, but as an adaptation of the previously existing sugar industry carrying with it some characteristics that are considered as a disadvantage, such as its seasonality (sugarcane is harvested in Brazil in the winter months, from May to October, meaning that it stops for about 5 months of the year). This represents a mill significantly larger (almost twice as large) implying more capital costs. Sugarcane is difficult to store (it is nearly 70% water, an inert material) and it needs to be processed not too long after its harvest (typically 24 hours after).

This genuine model, which allowed substantial gains for both consumers and mill owners, is giving sign of difficulties in expanding. Brazilian overall sugarcane ethanol production has stagnated at around 25 billion liters per year, and much of the difficulty is related to the incapacity to expand fuel ethanol production without producing additional sugar.

It was expected that as the oil companies, such as BP, Shell and Petrobras, entered in the sugar-ethanol sector in Brazil by acquiring existing sugar mills, they would go for dedicated energy or energy-chemical factories, concentrating their objectives to produce ethanol, electricity and may be chemical products. But unfortunately, these companies bought existing sugar-ethanol mills, maybe with the intention to learn about this market, but they did not innovate, remaining with the same production pattern they had inherited. It is true they are searching to innovate, particularly in producing 2G ethanol, but they do not abandon the sugar production, meaning that the "Brazilian model" still rules.

Successive crises (1989, 1999, 2010–15) indicate that besides difficulties associated with government lack of planning, there is an unseen or unrecognized crisis of the model itself. The flexibility levels have been built in a way to overcome momentary difficulties and were incorporated into the system. This, of course, was positive and gave "gas" to the PROÁLCOOL program in Brazil, despite the enormous planning difficulties many times associated with the young Brazilian democracy, notably the creation of the energy regulatory agencies such as ANP (National Agency of Petroleum, Natural Gas and Biofuels) and ANEEL (National Electricity Regulatory Agency) and the frequent instability they are victim to.

Therefore, it is not by accident that Brazilian ethanol production has stagnated for many years, if not decades, subordinating its expansion to the expansion of the sugar market or the technical capacity of Brazilian engines to operate with higher proportions for ethanol–gasoline blends. For instance, it is not by accident that the price difference between E100 and E27 at the gas station is "fixed" at the 70% level, indicating that it reflects the preconceived idea that E100 performs 25–30% less than E27 (Flex III) and the mills should try to maximize their benefits using Flex I strategies. Government, of course, acts when needed (Flex II mechanism).

But what about the future? Can Brazil conceive a model beyond its own? Should other sugarcane-producing countries copy the Brazilian model and should Brazil invent a new model, this time based on energy and chemical, disconnected from the food (sugar) and search new possibilities? The adoption of the Brazilian model of using low-quality sugar and molasses to produce ethanol probably would have positive consequences for the sugar market (smaller production but higher quality and higher prices benefiting less developed producing countries) and the almost immediate creation of an ethanol market, since around 15–20% of sucrose would be diverted to that end. However, this move involves political difficulties both at external and domestic levels.

HOW DID INNOVATION HELP THE EVOLUTION OF THE "BRAZILIAN MODEL"

Several authors have indicated how innovation has helped the Brazilian industry to achieve a leading position in both sugar and ethanol markets (Braunbeck et al., 2005; CTC, 2007; Cortez, 2010; Furtado et al., 2011; Braunbeck, 2013; Brito Cruz et al., 2014; Cortez, 2014b; Cortez et al., 2014, 2016). The important S&T contributions that can be mentioned are given in the following sections.

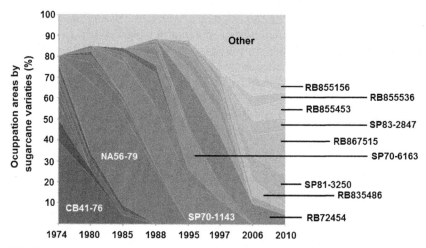

FIGURE 6.7 Increasing number of sugarcane varieties in Brazil. *Costa et al. (2011).*

Development of New Sugarcane Varieties

Brazil has conducted three important sugarcane breeding programs: the Agronomic Institute of Campinas—IAC; the RIDESA (former Planalsucar) program; and the CTC program. The varieties introduced by these programmes have significantly increased the sugarcane varieties planted in Brazil. In 1975, when PROÁLCOOL was launched there were only half a dozen varieties planted in Brazil as opposed to more than 500 planted today (Fig. 6.7). This does not only bring robustness and reliability on future harvests but also promotes increasing yields, having increased on average from 45–50 tonnes/ha to nearly 80–85 tonnes/ha today.

Other companies, such as Alellyx and Canaviallis (bought by Monsanto and terminated in 2015), were trying to introduce commercial varieties based on studies of cane genomics. These are not yet commercial varieties. More about sugarcane varieties in Brazil is available in Dinardo-Miranda et al. (2010) and Souza et al. (2015).

The Introduction of Green Cane Harvesting and Sugarcane Harvest Mechanization in Central-South Brazil

Another area in agriculture that went to a fast transformation in Brazil was the phase out of cane burning followed by the introduction of mechanization in the harvesting process. Until the mid-1990s sugarcane was all harvested burnt followed by hand cutting. However, since sugarcane plantations are concentrated in central-south Brazil, and more specifically close to urban areas, its burning prior to harvesting created some discomfort among the population which created the conditions for a protocol signed by the mills to phase out sugarcane burning.

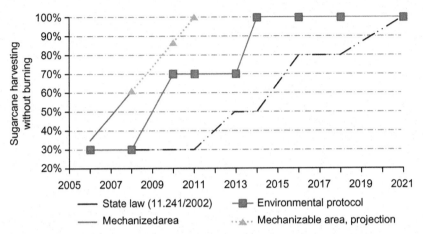

FIGURE 6.8   Evolution of sugarcane harvest mechanization in São Paulo State from 2001 to 2021. http://homologa.ambiente.sp.gov.br:80/etanolverde/resultado.asp.

The evolution of sugarcane burning phase out can be seen in Fig. 6.8. The law specified two moments for the burning phase out: 2017 and 2021, depending on the field slopes, which makes harvesting more difficult.

Today, practically all sugarcane fields are harvested without burning in São Paulo State and practically 70% all over Brazil. However, although green cane harvesting and mechanization can be seen as a positive contribution, it brings problems the sector did not have before. Table 6.1 presents some difficulties and what is been currently done to overcome these issues.

As can be seen from Table 6.1, this transition from burned harvesting to green cane mechanized harvest was not easy. The introduction of mechanization in the sugarcane harvesting process brought several problems, frequently related to inadequate harvesters. The utilized machines, regardless of their manufacturers, are very heavy and the combined effect, together with excessive operations, will result in important soil compaction. More about this is discussed below.

Evolution in Industrial Parameters of Ethanol Production

The already-mentioned synergy between sugar and ethanol is probably the main factor to be observed concerning the improvements in the industry. Besides, there were gains in the milling (extraction) sector and in the energy area with more efficient boilers and recovering part of now available trash for cogeneration (UNICA, 2008).

The combined effects of increasing agriculture (sugarcane yields) and industrial productivity can be better seen in Fig. 6.9 when these factors are aggregated into a "total productivity."

The present overall yield in the Brazilian sugarcane ethanol has stabilized at 7000 liters of ethanol/ha per year. This stagnation results from adverse climatic

TABLE 6.1 Comparison of Sugarcane Main Parameters Affected by Different Harvesting Possibilities: Burning With Manual Cutting and Green Cane Mechanized Harvesting

Parameter	With Burning Cane (Hand Cutting)	With Green Cane (Mechanical Harvesting)	What Is Being Done Currently
Sugar losses during harvesting	Order of 2–3% of sugar losses (visible and invisible losses) due to exudation	No losses by exudation	No need
	No losses because cane is cut in whole stalks	Order of 2–3% of sugar losses by chopping the cane in billets	Development of whole cane harvesting system
	Practically no damage since hand cutting is more precise	Damage to sugarcane "base" by harvester knives	Improve knives at the harvester
Soil compaction	Practically no compaction	Heavy traffic increase soil compaction decreasing cane yields	Development of ETC by CTBE for a minimum traffic field management
Trash in the field	No trash was left in the field because it was all burned	Green cane harvesting introduces trash, which is good but may bring more insects and more difficulties for the cane regrow	Development of partial trash removal from cane fields
Cane quality at the mill	Relatively clean, practically no dirt	3–5% of impurities brought with the cane to the mill	Development of cane cleaning systems at the mill

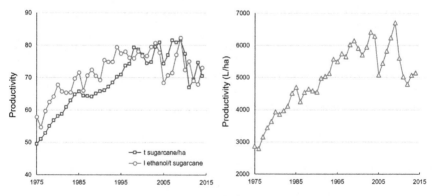

FIGURE 6.9 Agro-industrial productivity evolution of sugarcane ethanol through R&D from 1975 to 2015. *FAPESP BIOEN, Sep 23, 2009 and MAPA, 2015.*

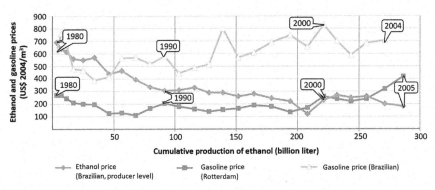

FIGURE 6.10   Sugarcane ethanol learning curve. *Goldemberg et al. (2008).*

conditions combined with malpractices in agriculture derived from difficulties to invest to maintain healthy sugarcane fields. In industry the main parameters (milling extraction and fermentation yield) somehow stagnated as a result of economics and also the limits of the technologies presently used.

However, the improvements can be better observed using the learning curve below. This curve somehow summarizes the learning process observed by the Brazilian sugar-ethanol industry (Fig. 6.10).

This exemplifies how unsubsidized sugarcane ethanol produced in Brazil was able to guarantee important profits and also lower prices of gasoline in Brazil. However, even though the domestic reputation was considered positive, internationally the opinion was not always the same.

HOW DID SUSTAINABILITY CRITICISM HELP THE ETHANOL INDUSTRY IN BRAZIL

As soon as biofuel production use started to become more important[1] worldwide, criticism greatly increased, particularly in the late 1990s and the new millenium. Several authors generalized some evidences, particularly for ethanol produced from grain feedstocks like corn in the United States and started to stablish correlations between higher prices of commodities and biofuel production. Today we know the high volatility in commodity prices is due to petroleum prices and not the production of biofuels.

The food versus fuel debate was a discussion many times dominated by passion since biofuels were pointed out as one of the causes of hunger in the world. It can be argued simply that hunger always existed, and that if biofuels stopped being produced, hunger would not be alleviated, but most likely the contrary would happen. More about food versus biofuels debate is available in Rosillo-Calle and Johnson (2010).

---

1. The world total agricultural area is around 1.5 billion ha. The area presently devoted to biofuels production is around 30 million ha.

Today, even the FAO (Dubois, 2014) recognizes that, if "done in the right way," meaning in a sustainable way, biofuel production not only doesn't represent a threat to food security but might also even enhance it.

From the social point of view, Moraes et al. (2011) present the social benefits derived from the production of sugarcane ethanol in Brazil, such as the creation of jobs and income. In one of the studies conducted, it is demonstrated that even children's schooling was improved in the Brazilian cities[2] where there is a sugar-ethanol mill.

Regarding the expansion of sugarcane, it had a major impact on land-use planning in Brazil. A few examples can be given: the Agro-Ecological Zoning of Sugarcane, prepared by the Ministry of Environment (MMA) (Fig. 6.11A), the Agroenvironmental Zoning for the cultivation of sugarcane, released on Sep. 18, 2008 by the government of the state of São Paulo (Fig. 6.11B) and based on the work of the BIOTA FAPESP, and the IAC work coordinated by Orivaldo Brunini in 2008, which resulted in the publishing of an agroclimatic suitability map for the state of São Paulo. In addition, the Brazilian Land Use Model (BLUM) (Nassar et al., 2011) can be mentioned as a contributor to the agricultural sector in Brazil.

Lessons learned from the biofuel sustainability debate include:

- Sustainability criteria for bioenergy is much more strict than presently used for food agriculture;
- Present technologies used in agricultural activities were developed considering food production in temperate climates, not bioenergy production needs in tropical climates;
- Technologies used in bioenergy crops are adaptations from the food sector, not tailored for bioenergy;
- If we want to make bioenergy sustainable we need to invent new sustainable technologies for bioenergy crops.

The debate should move from a "working on the problem" approach to a "working on the solution" approach. The focus should be on building a new agriculture, conceived from the beginning to satisfy the sustainability criteria. However, a few questions need to be asked at this point: do we have the idea of what we really want? Do we really know what the local and global correct indicators are? Do we really know what and how much they should be? Do we have sustainability targets for agriculture? How will sustainability, in its broad sense, induce new production models?

---

2. In the state of São Paulo, Brazil, where around 60% of sugarcane is produced in the country, there are around 150 sugar-ethanol mills. In a state with around 600 municipalities, this represents that nearly one every four cities in São Paulo State has a sugar-ethanol mill, which is usually the main taxpayer and job employer in that city.

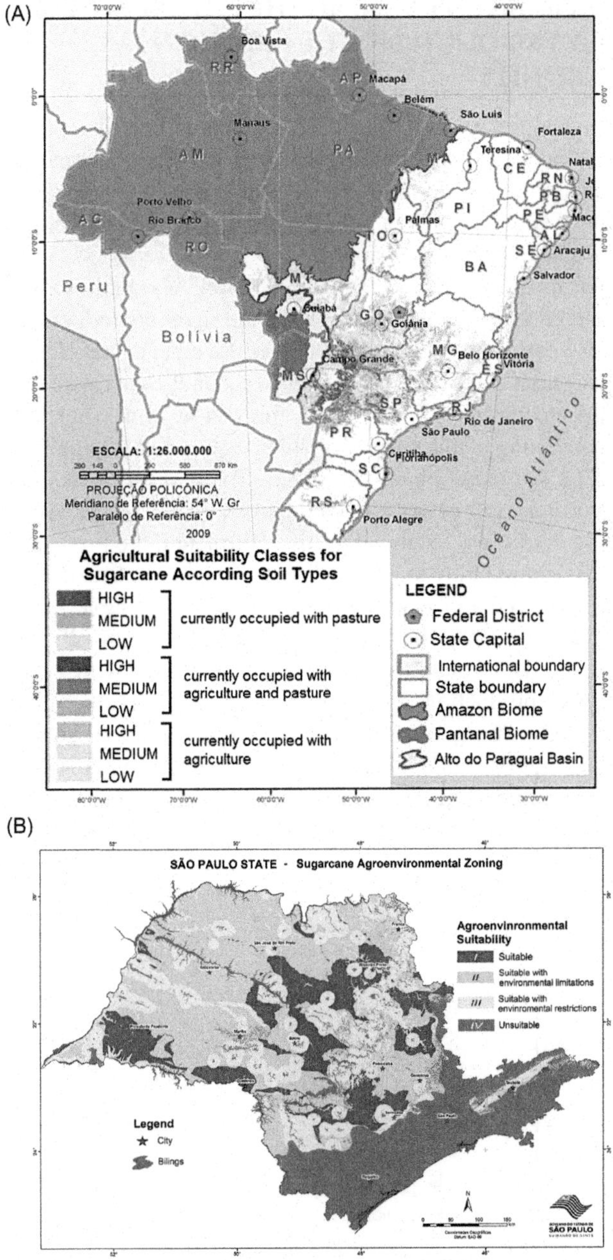

FIGURE 6.11 (A) Sugarcane agroecological zoning in Brazil. (B) Science-based sugarcane agroenvironmental zoning in São Paulo State. *For (A): EMBRAPA SOLOS (2009).* http://www. cnps.embrapa.br/zoneamento_cana_de_acucar/ZonCana.pdf. *For (B):* http://www.ambiente.sp.gov. br/etanolverde/zoneamento-agroambiental/.

FORECASTING AND PLANNING THE FUTURE SUGARCANE BIOENERGY PRODUCTION: THE CREATION OF "NEW MODELS"

In this session the possibilities to expand the existing Brazilian sugarcane ethanol model are analyzed, highlighting its challenges and prospects of growth and diversification. First, it is recognized that although the present "Brazilian Model" allowed Brazil to build a considerably large bioenergy program, it represents the obstacle for its expansion.

As was discussed before, the "Brazilian Model" was built with sugar as the main product and ethanol and electricity as the main coproducts. Most of the trials to build "energy mills," dedicated to produce ethanol and electricity, failed because sugar presents more economic returns than the energy products, where profit margins are slimmer. For instance, this can be seen when analyzing the decline or abandoning of autonomous distilleries concept originally built during the PROALCOOL second phase (1979–82). Today most of the mills (around 90%) produce sugar and ethanol. Even the mills owned by traditional oil companies such as Petrobras, Shell and British Petroleum, are producing sugar. In fact, Shell is today probably the largest sugar producer in Brazil.

Therefore, when ethanol expansion is analyzed, for example, taking the Leite (2009), Leite et al. (2009) study and EMBRAPA SOLOS (2009) projection (Fig. 6.12), it is obvious that this can not be accomplished with a correspondent increase in sugar production. Taking the case of EMBRAPA SOLOS (2009), if sugarcane production double by 2025, this would imply a great increase in sugar demand, requiring another China to absorb all this extra sugar supply.

There is, in fact, enough land in Brazil for a significant increase in bioenergy production, such as sugarcane ethanol and bioelectricity. EMBRAPA SOLOS

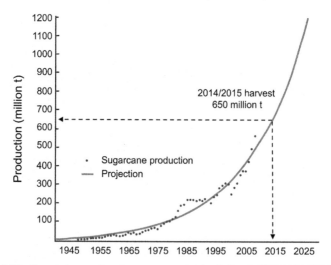

FIGURE 6.12   Expected expansion of sugarcane production in Brazil. *EMBRAPA SOLOS (2009).*

(2009) estimates in the agroecological zoning that 60 million ha are considered adequate for sugarcane production in Brazil. This represents more than six times what is produced today. However, a new model, not dependent on sugar, needs to be developed and implemented, but how?

Each production unit of the present model (sugar-ethanol distillery) consists also of one mill and a sugarcane plantation around it. The largest mill in Brazil crushes around 8 million tons of cane, utilizing around 100,000 ha in sugarcane cultivation. In the present sugar-ethanol model this represents around 3.6 million liters of ethanol per day (22,400 barrels of ethanol per day or around 402,500 barrels of ethanol per year), considering 180 days cane season. Presently exists around 150 sugar-ethanol mills located in the state of São Paulo. All of them combined produce around 15-20 billion liters of ethanol per year (around 94-126 million barrels of ethanol per year). So, the ethanol produced in the State of São Paulo can be compared to, for example, Paulínia, the largest oil refinery in Brazil, responsible for about one-fifth of all oil consumed in Brazil, with a refining capacity around 400,000 barrels of oil per day, or 146 million barrels of oil per year. Just for comparison, the largest sugar-ethanol mill in Brazil produces in volume 22,400 barrels of ethanol per day, meaning 5.6% of oil processed by Paulínia.

There is a need to create new production model(s), dissociated from sugar production, but what features would the new model(s) need to have?

Inventing a New Sustainable Agriculture for Sugarcane Bioenergy

A new sustainable agriculture would have to be developed considering the following aspects:

1. A new feedstock: "energy cane," focus on energy not on sucrose;
2. To build soil fertility, not promote soil mining. Promote mulching, back soil and introduce precision agriculture;
3. New soil conservation practices for sugarcane: no-till planting, straw blanck-eting, 80% less compaction;
4. Low use of agrochemicals through development of technologies such as biological control and transgenic varieties;
5. New agricultural machinery with lower cost, lower weight, lower losses and higher operating capacity;
6. Landscape improvement: promote water conservation, and biodiversity conservation through establishment of ecological corridors.

An example of this new agriculture development to meet the requirements of strict sustainability criteria imposed for bioenergy is the "Minimum Impact Agriculture" currently developed by the Brazilian Ethanol S&T National Laboratory—CTBE (Braunbeck, 2013). First, it is recognized that around 60–70% of sugarcane ethanol costs derives from the raw material (sugarcane) itself. In addition, three operations in cane harvesting: cutting, loading and transport are responsible for 40–50% of total cane costs at the mill. It was concluded

FIGURE 6.13 Minimum impact agriculture—the CTBE project. *Braunbeck (2011).*

that sugarcane harvesting is too costly and the harvesters are too heavy and therefore compact too much on the soil, decreasing cane yields.

In response, the CTBE research group is developing the ETC "traffic control structure," using a totally new concept for sugarcane: instead of using conventional large tires, the ETC uses narrow tires that will only pass on specified previously compacted tracks freeing the sugarcane cultivated zone from any king of compaction. Besides, the ETC does all field operations being also responsible for sugarcane planting, besides harvesting itself. Fig. 6.13 illustrates the ETC.

Besides the new aspects introduced by CTBE's ETC technology, other important issues still need to be overcome such as the elimination of the current intermittent feedstock supply, which is limited by the sugarcane season. This represents a major obstacle to the present model and needs to be addressed either by introducing complementary feedstocks, suck as sugarcane fiber for 2G ethanol, or materials that can be easily stored, such as cassava or even corn or sorghum.

New Industrial Production Models Integrated With the Oil, Electricity, Chemical and Other Related Industries

New industrial production models would also need to be developed. There is today a firm decision of the corn ethanol industry in the United States and the sugarcane ethanol industry in Brazil to develop so-called second-generation ethanol (2G) from lignocellulosic materials. Several projects are currently ongoing, even on a pilot scale, to demonstrate its technical and economic feasibility. In Brazil, the most important is being developed by GranBio in the state of Alagoas in the Brazilian northeast (GranBio, 2014). If successful this project could open the doors for new possibilities for ethanol expansion without sugar production.

Another strategy utilizes the pyrolysis technology to convert the whole sugarcane into products, not necessarily ethanol. Laboratory tests conducted at UNICAMP, Brazil, utilizing whole sugarcane (not only bagasse or trash) open a new possibility in this field. Converting the whole cane resources implies not adopting an adaptation solution such as when bagasse or trash is utilized. Instead, the whole cane, with all its sucrose and fibers are chopped and partly dried prior to the pyrolysis process. Therefore, all the sucrose present in the cane is submitted to pyrolysis and no sugar or even ethanol is produced. In this process, the obtained products are bio-oil (a liquid-like material such as "tar"), a gas (basically composed by CO, CH_4, H_2 and CO_2), and charcoal. The bio-oil and charcoal can be easily transported, for example to an oil refinery and the gas can be burned for the sugarcane drying process. This concept can be utilized to eliminate the conventional sugar-ethanol mill directly connecting the sugarcane to the oil refinery.

Other production models can be proposed. The main idea is to overcome transportation difficulties which usually feature biomass transportation. For example, today there is an increasing market for biomass pellets in Europe. Maybe this can be another possibility for sugarcane bioenergy, not necessarily related to sugar-ethanol.

CONCLUSIONS ON THE PROSPECTS OF THE BIOETHANOL INDUSTRY IN BRAZIL AND ITS PRODUCTIVE MODEL

There is enough land in the world to expand the sustainable production of biofuels. In particular, Brazil presents exceptional conditions of land availability and climate to produce both bioenergy and food, without threatening its biodiversity and ecological sanctuaries.

On the other hand, it can be stated that the future of the bioethanol industry will be ruled by the two issues: perception about global warming and its related climate risks and oil prices and the relative prices of fossil and biofuels. In the last decades these have been the important drives and it is expected they will still be in the 21st century.

In the last 40 years, Brazil has developed a very genuine production and utilization model for sugarcane ethanol allowing substantial results both in energy and social-economic aspects. However, the present sugar-ethanol model presents important limitations, the most important being the food–energy interdependence, which needs to be overcome if a substantial expansion is to be expected. New sustainable production models, free from self-imposing limitations, should be conceived. This will imply significant changes in bioenergy agriculture and also on the business model.

As a main conclusion, it can be stated that bioenergy economic and environment learning curves can and also should walk together. For this to happen, the new sustainable agriculture needs to be invented tailored to address the sustainability criteria. Also, the entire production chain, developed for specific markets, needs to be created, oriented to the partial substitution of oil, but in conjunction with the oil industry.

Finally, bioenergy may play a much more important role in the future but old production patterns need to be replaced by a more scientific approach and new business models.

ACKNOWLEDGMENT

The authors would like to thank the São Paulo State Research Foundation (FAPESP) for the necessary funds to conduct this research and publication.

REFERENCES

BNDES, 2008. Bioetanol de cana-de-açúcar: energia para o desenvolvimento sustentável (in Portuguese) organized by BNDES and CGEE—Rio de Janeiro, 319 pp.

Boeing, Embraer, and FAPESP, June 2013. Flightpath to Aviation Biofuels in Brazil: Action Plan Report, São Paulo. http://www.fapesp.br/publicacoes/flightpath-to-aviation-biofuels-in-brazil-action-plan.pdf.

Braunbeck, O., "Máquina versátil" (in Portuguese) Access in April 2015. http://revistapesquisa.fapesp.br/2011/11/11/maquina-versatil/.

Braunbeck, O., 2011. Máquina versátil. Available at: <http://revistapesquisa.fapesp.br/2011/11/11/maquina-versatil/>. Access in April 2015.

Braunbeck, O., 2013. Presentation to BNDES. http://www.stab.org.br/palestra_sistematizacao_2013/06_oscar_braunbeck_23.pdf.

Braunbeck, O., Macedo, I.C., Cortez, L.A.B., 2005. Modernizing cane production to enhance the biomass base in Brazil Chapter 6. In: Silveira, S. (Ed.), Bioenergy—Realizing the Potential Elsevier, Oxford, pp. 75–94.

Brito Cruz, C.H., Cortez, L.A.B., Souza, G.M., 2014. Biofuels for transport Future Energy: Improved, Sustainable and Clean Options for Our Planet. Elsevier, London, 716 pp., ISBN 9780080994246, http://find.lib.uts.edu.au/;jsessionid=B944BE4C8E2E60E5C1BDB7274F25 240F?R=OPAC_b2845998.

CGEE, Sustainability of Sugarcane Bioenergy. Brasília, DF: Center for Strategic Studies and Management (CGEE), 336 pp., 2012.

Copersucar, 1989. Proálcool Fundamentos e Perspectivas. Cooperativa de Produtores de Cana, Açúcar e Álcool do Estado de São Paulo Ltda., São Paulo, 121 pp.

Cortez, L.A.B. (Ed.), 2010. Sugarcane Bioethanol: R&D for Productivity and Sustainability Editora Edgard Blucher, São Paulo. 992 pp., ISBN 978-85-212-0530-2.

Cortez, L.A.B. (Coord.) 2014a. Roadmap for Sustainable Biofuels for Aviation in Brazil Coordenador, São Paulo, 272 pp.

Cortez, L., 2014b. Shaping New Agriculture for Bioenergy. Presentation at the Workshop on Biofuels and Food Security Interactions. IFPRI, Washington, DC, November 20, 2014.

Cortez, L.A.B., Souza, G.M., Cruz, C.H.B., Maciel, R., 2014. An assessment on Brazilian Government initiatives and policies for the promotion of biofuels through research, commercialization and private investment support. In: da Silva, S.S., Chandel, A.K. (Eds.), Biofuels in Brazil Springer, Cham. 29 pp. <http://www.eolss.net/sample-chapters/c08/E6-185-21.pdf>, <http://www.springer.com/life+sciences/microbiology/book/978-3-319-05019-5>.

Cortez et al. (2016) "Universidades e Empresas: 40 anos de Ciência e Tecnologia para o Etanol Brasileiro" Editora Blucher, São Paulo (in press), 2016.

Costa, M.D.-B.L., Hotta, C.T., Carneiro, M.S., Chapola, R.G., Hoffmann, H.P., Garcia, A.A.F., et al., 2011. Sugarcane Improvement: how far can we go? Curr. Opin. Biotechnol. 23, 1–6.

CTC, 2007. Communication obtained at Centro de Tecnologia Canavieira (Sugarcane Technology Center), Piracicaba, SP.

Datagro, 2008. Based on UNICA and ICONE.

de Moraes, M.A.F.D., Zilberman, D., 2014. Production of Ethanol From Sugarcane in Brazil: From State Intervention to a Free Market. Springer, Cham, 221 pp.

Dinardo-Miranda, L.L.; Vasconcelos, A.C.M.; Landell, M.G.A., 2010. Cana-de-Acucar. Agronomic Institute of Campinas-IAC, 882 pp. (in Portuguese).

Dubois, O., 2014. Presentation at the Workshop on Biofuels and Food Security Interactions. IFPRI, Washington, DC, November 20, 2014.

EMBRAPA SOLOS, 2009. Zoneamento Agroecológico da Cana-de-açúcar: Expandir a produção, preservar a vida, garantir o futuro. Empresa Brasileira de Pesquisa Agropecuária – Embrapa: Rio de Janeiro, 58 p. Available at: <http://www.mma.gov.br/estruturas/182/_arquivos/zaecana_doc_182.pdf>.

FAPESP, 2007. Brazil world leader in knowledge and technology of sugarcane ethanol, São Paulo, 75 pp.

FAPRI, 2008. U.S and World Agricultural Outlook. Food and Agricultural Policy Research Institute-FAPRI, Iowa State University and University of Missouri-Columbia, Ames, Iowa395. Available at: <http://www.fapri.iastate.edu/outlook/>.

FAPRI, 2012. U.S and World Agricultural Outlook – Section Biofuels. Food and Agricultural Policy Research Institute-FAPRI, Ames. Available at: <http://www.fapri.iastate.edu/outlook/2012/tables/5-Biofuels.pdf>.

Finguerut, J., 2005. Simultaneous production of sugar and alcohol from sugarcane. Proceedings of the XXV International Society of Sugarcane Technologists - ISSCT Congress, Guatemala City, Guatemala, January 30 to February 4, Vol. 25, pp. 315–318.

Fulton, L.M., 2013. The Need for Biofuels to Meet Global Sustainability Targets. Presentation at the BIOEN-BIOTA-PFPMCG-SCOPE Joint Workshop on Biofuel & Sustainability. FAPESP, São Paulo, February 26, 2013. Available at: <http://www.fapesp.br/eventos/2013/02/BIOEN-BIOTA/Lewis_Fulton.pdf>.

Furtado, A.T., Scandiffio, M.I.G., Cortez, L.A.B., January 2011. The Brazilian sugarcane innovation system. Energy Policy 39 (1), 156–166.

Goldemberg, J., Coelho, S.T., Nastari, P.M., Lucon, O., March 2004. Ethanol learning curve—the Brazilian experience. Biomass Bioenergy 26 (3), 301–304.

GranBio, 2014. GranBio inicia produção de etanol de segunda geração (in Portuguese). http://www.granbio.com.br/wp-content/uploads/2014/09/partida_portugues.pdf.

Leite, R.C.C. (Ed.), 2009. Bioetanol Combustível: uma oportunidade para o Brasil Centro de Gestão de Assuntos Estratégicos-CGEE, Brasília. 536 pp., (in Portuguese).

Leite, R.C.C., Leal, M.R.L.V., Cortez, L.A.B., Griffin, W.M., Scandiffio, M.I.G., 2009. Can Brazil Replace 5% of the 2025 Gasoline World Demand with Ethanol. Energy 34, 655–661.

MAPA, 2015. Anuário Estatístico da Agroenergia 2014: Statistical Yearbook of Agrienergy 2014. Ministry of Agriculture, Livestock and Supply-MAPA and Secretary of Production and Agrienergy. Brasília: MAPA/ACS, p. 205. Available at: <http://www.agricultura.gov.br/arq_editor/anuario_agroenergia_WEB_small.pdf>.

MME-EPE, 2007. Matriz Energética Nacional 2030. Ministério de Minas e Energia-MME, Empresa Brasileira de Pesquisa Energética-EPE, Brasília, 254 p, 2007. Available at: <http://www. mme.gov.br/documents/10584/1432020/Matriz+Energ%C3%A9tica+Brasileira+2030+- +(PDF)/708f3bd7-f3ed-4206-a855-44f6d4db29f6;jsessionid=E522BB848EBC527D623B696 E5C804D4F.srv154>.

Moraes, M.A.F.D., et al., 2011. Ethanol and Bioelectricity. Chapter 2: Social Externalities of Fuels UNICA, São Paulo, Available from: file:///C:/Users/luis.cortez/Downloads/089b4fc3264d69e8 05b1514cccf2d53d%20(1).pdf.

Nassar, A.M., 2011. Simulating land use and agriculture expansion in Brazil: food, energy, agro-industrial and environmental impacts". Icone, São Paulo, 2011. Relatório científico final. Available at: <http://www.iconebrasil.com.br/datafiles/publicacoes/artigos/2002/simulating_ land_use_and_agriculture_expansion_in_brazil_0902.pdf>.

Rosillo-Calle, F., Cortez, L.A.B., 1998. Towards Proalcool II—a review of the Brazilian bioethanol programme. Biomass Bioenergy 14 (2), 115–124.

Rosillo-Calle, F., Johnson, F.X. (Eds.), 2010. Food Versus Biofuels Zed Books, London and New York, NY. 232 pp.

Souza, G.M. Victoria, R. Joly, C. Verdade, L. (Eds.), (2015). Bioenergy & Sustainability: Bridging the Gaps, vol. 72 SCOPE Report, Paris. ISBN 978-2-9545557-0-6

UNICA, 2008. Sugarcane Industry in Brazil: Ethanol, Sugar, Bioelectricity. From: www.unica.org.br.

Chapter 7

Ethanol Production Technologies in the US: Status and Future Developments

B.A. Saville[1], W.M. Griffin[2] and H.L. MacLean[3]
[1]Department of Chemical Engineering and Applied Chemistry, University of Toronto, Toronto, ON, Canada [2]Department of Engineering and Public Policy, Carnegie Mellon University, Pittsburgh, PA, United States [3]Department of Civil Engineering, University of Toronto, Toronto, ON, Canada

INTRODUCTION

As of August 2015, there were 215 operating ethanol plants in the US using starch/sugar as the feedstock, representing 15.5 billion gallons of production capacity (US Renewable Fuels Association (RFA), 2015; Ethanol Producer Magazine, 2015). The majority of these plants are located in the US Midwest "corn belt," and use corn for feedstock; about 10% use sorghum when corn is not available. A handful of plants use sugars from adjacent beverage operations. Approximately 2 billion gallons of production capacity is collocated with wet milling operations that produce a varied product slate that includes starch, sweeteners, fiber and a variety of other food/feed products. The balance of the production is via dry milling operations, most of which produce distillers' dried grains with solubles, either in wet or dry form (WDGS or DDGS, respectively).

Currently, two commercial-scale cellulosic ethanol plants are operating in the US, POET-DSM and Abengoa, with others in the commissioning and start-up phase. Dupont also has a commercial-scale plant nearing completion, with commissioning and start-up imminent. Collectively, these three plants represent 80 million gallons/year of production capacity. All three plants use a biochemical conversion technology and crop residues as the feedstock. Abengoa uses wheat straw, while POET-DSM and Dupont use corn stover. Two other facilities (Ineos Bio and Fiberight) are at the technology demonstration scale designed to produce between 5 and 10 million gallons/year and use municipal solid waste or vegetative waste as feedstocks.

Quad County cellulosic ethanol produces 2 million gallons/year of ethanol using corn fiber. Another 8–10 plants would be classified as pilot facilities, with production capacities typically less than 1 million gallons/year, and usually only operated on an intermittent basis. These facilities use a variety of feedstocks, including sugarcane bagasse, grasses, hardwoods and energy beets. Another nine plants have been proposed, with capacities between 5 and 100 million gallons/year, representing an aggregate production capacity of 233 million gallons/year. Notably, many of these proposed plants would be located outside the corn belt, and use feedstocks other than agricultural residues.

FIRST-GENERATION TECHNOLOGY

Feedstock Choices: Corn Versus Other Grains

The primary feedstock used for first-generation ethanol production in the US is #2 yellow dent corn (or field corn), which has a very thick outer skin that renders it unsuitable for human consumption, although it can be consumed as animal feed. This industrial-grade grain is also used to make starch and sweeteners such as glucose and high-fructose corn syrup. This variety of corn contains about 70% starch, along with about 8–11% protein, 3–4% oil and about 10% fiber (Huang et al., 2012).

The technology to convert field corn into ethanol can be readily applied to other starch-containing feedstocks, with minor adaptations to account for differences in starch, protein and β-glucan content. Sorghum is commonly used as an alternate feedstock for plants in Kansas and Texas; high-starch, low-protein wheat is used in several Canadian and international plants.

Dry Milling

Dry milling (Fig. 7.1) involves grinding the incoming grain, then processing it through a series of steps to liquefy the flour and generate fermentable sugars. Amylases are added at two points in the process—the initial slurry step, and the liquefaction step, which follows a jet cooking operation that uses high-temperature steam to swell the starch. Following liquefaction, the slurry is fed to a batch fermentation system, where glucoamylase and yeast are also added. The typical fermentation time is 42–55 h. Multiple fermenters are used to facilitate batch operation of this step, with the fermentation cycle time including a clean-in-place step prior to the addition of fresh mash that commences the start of the fermentation process. Final ethanol titers between 14 and 18 wt% are typical. Once the fermentation is complete, the ethanol-laden mash is transferred to a beer well that ultimately feeds the first stage of a two-stage distillation system. The first distillation stage includes all of the unconverted solids, which are recovered at the bottom of the column, while the overhead, typically containing about 30–40% ethanol, is fed to a second

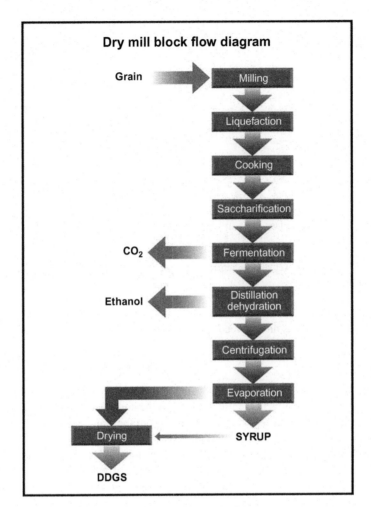

FIGURE 7.1 Block flow diagram: dry mill ethanol production.

distillation column that purifies the ethanol to a concentration near its azeotrope (about 190 proof). The hydrous ethanol is then dehydrated using a set of molecular sieves, producing a 99.5% ethanol product. This product is then denatured to meet government regulations.

The wet solids recovered at the bottom of the first distillation column are centrifuged, with the liquid sent to a set of multiple effect evaporators to recover water for reuse in the process (typically added to the first slurry reactor), while the solubles (mainly sugars) are typically blended with the wet distillers' grains to produce wet distillers' grains with solubles (WDGS). The WDGS may be sold as is to nearby feedlots due to the limited shelf life of the wet product, or

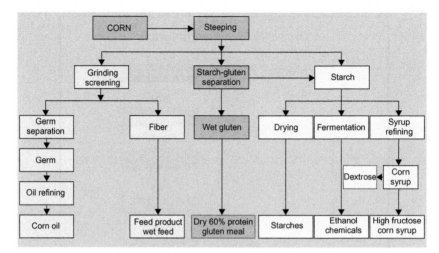

FIGURE 7.2 Ethanol wet milling process. *Adapted from* http://indianaethanolproducers.org/wet_milling.html.

optionally dried to produce DDGS, a stable, protein-rich product that can be shipped worldwide as animal feed.

Wet Milling

The wet milling process (Fig. 7.2) is distinctly different from the dry milling process. Wet milling begins by gentle steeping of the corn, which allows germ, oil, gluten and fiber to be recovered and isolated from the starch. The starch can be dried and sold, hydrolyzed into glucose syrup (and optionally further converted into high-fructose corn syrup), or hydrolyzed using amylase and glucoamylase and fermented into ethanol. The wet mill process to produce ethanol is distinct in the diversity of high-value coproducts, but also, in the fact that distiller's grains are not produced. The removal of solids prior to hydrolysis and fermentation means that the distillation process is simpler and more energy-efficient.

Recent Technological Advances, Efficiency Improvements and New Coproducts

Various technological improvements have been researched, and some commercialized, to improve upon the basic grain to ethanol process. Examples include:

1. The cold-cook (BPX) technology developed and commercialized by POET, which avoids the high-temperature jet cooking process to save thermal energy, while employing a more aggressive milling process and additional/different amylase to break down the starch. Another key advantage of this process is that it avoids the enzyme inactivation that typically occurs in a high-temperature jet cook operation. Thus, the enzymes added to the slurry

tank remain active, and the second enzyme addition to the liquefaction system is avoided or reduced.

2. Extraction of crude corn oil has been implemented in many dry mill plants (Greenshift; US Grains Council; Jessen, 2014). The quality of the oil is lower than that of oil extracted via wet milling, but is still suitable as a low-cost feedstock for biodiesel production. Another advantage is that oil removal enhances the protein content (and value) of the distillers' grains.

3. Technologies such as Valicor's modular VFRAC process and ICM's FST process have been developed for dry fractionation of grain components to recover, for example, gluten or fiber prior to the normal dry milling process (Jessen, 2014; ICM Inc.). These fractionated products are typically food grade, and thus have high value, which compensates for the additional capital and energy costs of the extraction process. ICM's process removes 2 kg of fiber from 25 kg of corn, leading to higher sugar concentrations from hydrolysis, and higher ethanol titers during fermentation, which reduce energy demand during distillation. These processes can also enhance plant throughput.

SECOND-GENERATION TECHNOLOGY

Role of Feedstocks and Feedstock Preprocessing

A key driver of interest in second-generation ethanol production is the fact that there are many more feedstock options available, grown in more diverse regions outside the typical agricultural zones. Potential feedstocks can be divided into several categories:

1. Agricultural residues such as corn stover, cane bagasse, and straw from wheat, rice or barley production;
2. Woody biomass, including "soft" hardwoods unsuitable for pulp/paper applications, forest residues, softwoods and tree plantations targeted at bioenergy production, such as willow and hybrid poplar;
3. Dedicated energy crops, such as switchgrass, *Miscanthus*, biomass sorghum and energy cane.

Crop residues are of interest because they are already grown, but often left in the field, plowed into the soil, or harvested for other low-value uses. In the case of most grain crops, residues such as stover or wheat straw are produced in nearly equal amounts as the grain itself. However, the equipment and infrastructure for harvest and storage of residue have only recently been developed, and are key factors influencing the delivered cost of the residues to the biofuels plant. Other critical factors include determination of the fraction of the residues that can be sustainably removed from the soil without adverse effects on soil carbon, storage conditions that limit degradation of the harvested residues, risks associated with fire or other loss during long-term storage, and the potential value for densification of the residues, which otherwise have a low free-fall bulk density.

A particular challenge is the short residue harvesting time and the need for several months storage in anticipation of year-round biofuel production. Typical yields of residues are on the order of 4–8 dry tonnes/ha, but the "harvestable amount" could be as little as 25% of these values. For example, POET-DSM reports that they will harvest only 20–25% of the available corn stover for its Project Liberty facility in Emmetsburg, Iowa (POET-DSM).

Cane bagasse, such as that available in Florida, Louisiana and Texas, is distinctly different from other agricultural residues, because whole cane stalks are delivered directly to the biofuel facility, and the bagasse is an on-site byproduct of sugar production or the production of ethanol from sucrose in the cane. Conversion of bagasse is not faced with the same harvest/logistics issues as other crop residues, but there can still be issues with degradation during storage outside of the regular harvest period for cane. There is, however, a direct parallel between collection of cane tops and leaves and the harvesting/collection of agricultural residues from grain crops.

Woody biomass is abundant but, in many cases, is more difficult to access economically or more valuable for other uses (paper, lumber, etc.). Thus, for biofuels, the focus is upon lower-value woody materials, such as willow or poplar, that are still relatively abundant but less desirable for pulp, paper and wood products. A key advantage of woody biomass feedstocks is the ability to harvest 10–11 months/year. This "direct from field" harvest can dramatically reduce or even eliminate the logistics and storage issues encountered with most agricultural residues. Since trees are multiyear crops, they are also less prone to one-off losses due to crop failures that may occur with annual grain crops. At the same time, this also means that the trees must be planted several years in advance in order to be readily available for a bioenergy facility. Harvesting equipment for woody biomass is readily available, and logistics are well established.

Dedicated energy crops show significant promise as feedstocks because of (1) their potential to be grown in more geographically dispersed areas compared to conventional crops and (2) their potential for much higher yields relative to crop residues, which should simplify harvesting and transport logistics, improve land use metrics and reduce costs. In North America, particular attention has been paid to switchgrass, biomass sorghum, *Miscanthus* and energy cane. Plantations of hybrid poplar and willow would also be classified in this category.

Day (2015) measured energy cane yields between 9 and 16 dry tonnes/acre in Louisiana, depending upon the variety. The Canergy group has obtained energy cane yields of 16–30 dry tonnes/acre when grown in southern California (Brummels and Saville, 2015). Day (2015) also reported sorghum yields from 1–9 dry tonnes/acre, while field trials at Texas A&M led to average yields between 3 and 7 dry tonnes/acre, depending on site and phenotype (Gill et al., 2014). Wullschleger et al. (2010) reported on switchgrass yield profiles based upon 39 field trials across the US, and noted that in upland regions, the switchgrass yield was 3.9± 1.9 dry tonnes/acre, versus 5.8± 2.6 dry tonnes/acre in lowland regions. Most field trials reported yields between 4.5 and 6.2 dry tonnes/acre. Select lowland ecotypes such as Alamo, planted in Texas, Alabama and Oklahoma, tended to have the highest yields, exceeding 12 dry tonnes/acre. Yield was sensitive to cultivar, soil type, climate (temperature and moisture) and amount of N fertilizer applied (although more N did not necessarily result in a higher yield).

While it is unlikely that a single type of dedicated energy crop will be suitable across all climatic regions, it is likely that some type of dedicated energy crop could be grown in any given region, albeit with differences in yields and nutrient requirements. These differences could materially impact the plant-gate delivered feedstock cost, and also the life cycle greenhouse gas (GHG) emissions, which will be affected by farming energy for planting and harvest (and thus, yields), and the carbon intensity of the agricultural inputs. It is important to note that some of these dedicated energy feedstocks face the same harvest/storage/logistics issues as agricultural residues, while others can be harvested over 6–10 months, which can provide a significant advantage.

A challenging/confounding issue with all second-generation feedstocks is their moisture content at harvest, and their free-fall bulk density. Woody biomass materials tend to have an equilibrium moisture content between 40 and 60%, whereas agricultural residues tend to contain 10–20% moisture. Dedicated energy crops can be fairly dry or have moisture levels comparable to those for harvested wood. On the surface, one would assume that a dry feedstock is beneficial. However, these types of feedstocks suffer from two key disadvantages: They tend to have a low free-fall bulk density, which means shipping and storage are dictated by volume rather than weight; and also more trucks are required, along with more storage space. However, dry materials also tend to be more amenable to long-term storage, since they are less prone to degradation. This also elevates the risk of loss by fire and the risk of dust explosions. Dry material, especially overdried material, also tends to be susceptible to collapse of the internal fibrils, rendering pretreatment and hydrolysis more difficult, while increasing enzyme requirements.

By comparison, materials containing higher levels of water are easier and typically more cost-effective to transport because the trucks can be loaded to road limits. The main disadvantage is that they are more prone to degradation, and need to be used within a short time, rather than stored long term. They do have the advantage of easier, more consistent pretreatment and hydrolysis because the bound water present ensures more even heat transfer, and prevents fiber collapse that would otherwise create challenges during hydrolysis. Furthermore, process water demand is less, because the feedstock supplies a large fraction of the water required for hydrolysis (or pretreatment, if a liquid-phase pretreatment is selected). Owing to their higher free-fall bulk density, a "wet" feedstock will also require less capital investment for feedstock handling at the plant, because these unit operations are mainly sized based upon volume, rather than mass.

In an effort to overcome some of the significant challenges with low bulk density feedstocks, researchers have begun to examine densification methods, either pre- or post-pretreatment (Kim and Dale, 2015). These densification steps can overcome many of the transport, storage and handling issues associated with processing low bulk density feedstocks, but they cannot overcome (and may worsen) the issues associated with fiber collapse that render pretreatment and hydrolysis more difficult. If densification is performed after pretreatment, and pretreated materials are stored in depots for ultimate shipment to a larger facility, there are still issues with additional capital required for feedstock handling prior to pretreatment, and one may lose opportunities for heat integration between the pretreatment system and downstream unit operations that could use the lower-grade heat emitted from the pretreatment process. Ultimately, much more research is required in this area in order to fully assess the tradeoffs involved in use of densified biomass.

Pretreatment Options

In spite of a large number of laboratory studies of biomass pretreatment, the process is relatively poorly understood because laboratory conditions rarely mimic commercial-scale operation. This is an important aspect when considering literature on pretreatment. In this chapter, we mainly focus on operational aspects of pretreatment processes that have been employed at a pilot or commercial scale. The reader can consult several reviews and book chapters on pretreatment for additional information on mechanistic elements of biomass pretreatment (Sun and Cheng, 2002; Mosier et al., 2005; Chandra et al., 2007; Galbe and Zacchi, 2007; Hu et al., 2008; Sanchez and Cardona, 2008; Saville, 2011).

The primary goal of pretreatment is to break bonds between cellulose, hemicellulose and lignin, and to activate cellulose, facilitating its hydrolysis (Saville, 2011). Several methods have been proposed to activate cellulose, which also tend to (partially) solubilize lignin and hemicellulose. Pretreatment is based upon a series of mechanical steps and chemical reactions, which may or may not include exogenous catalysts to facilitate the reactions. Particle size and size distribution are important to ensure efficient, uniform heat and mass transfer; this is generally accomplished via a front end size reduction, refining or milling operation. This is followed by the actual pretreatment process, which may include a presoaking step to embed catalyst such as acid or base inside the fiber structure. The pretreatment process generally takes place at an elevated temperature; typically 180–220°C is needed if using acids, steam or hot water, but lower temperatures are feasible if using an alkaline catalyst such as ammonia or caustic.

Among the many pretreatment methods proposed, current commercial- and pilot-scale plants mainly use (1) autohydrolysis, either as high-temperature hot water or steam, (2) dilute sulfuric acid in combination with steam or (3) aqueous ammonia pretreatment. Time will tell which methodology enjoys long-term commercial success. If an acid or base catalyst is used, it generally must be neutralized or washed off the pretreated biomass prior to hydrolysis. If the catalyst is expensive, hazardous, supplied in excess or in vapor form (such as ammonia), it typically needs to be recovered for re-use. These steps add to processing cost, and affect energy use and GHG emissions, and thus must be balanced against improvements in cellulose reactivity that can be achieved when these catalysts are used. Addition of catalysts also often limits options for heat integration because of the presence of contaminants in the solvent or vapor streams. In contrast, exhaust steam from a steam explosion process can supply heat to downstream distillation or evaporation operations. Thus, there is a general perception that better energy efficiency and lower capital costs can be obtained with autohydrolysis pretreatments compared to systems using acids that require corrosion-resistant metals, or alkaline systems that require neutralization and/or catalyst recovery.

Enzyme Hydrolysis

The efficacy of enzyme hydrolysis is highly dependent upon the pretreatment process, which dictates the reactivity of the cellulose, ease of solubilization and hydrolysis of hemicellulose, and the particle size distribution of the feed to the hydrolysis reactor. High solids hydrolysis is key to reducing capital costs, but can create rheological challenges that adversely affect mixing and require high power input to ensure sufficient agitation. Enzymes play a key role in viscosity reduction during hydrolysis, converting insoluble solids into soluble oligomers and monomers that are easier to mix (Di Risio et al., 2011). Conventional batch hydrolysis is particularly challenging at large scale due to issues with mixing; semibatch or continuous hydrolysis can overcome these inherent issues, but may extend the duration of hydrolysis or increase susceptibility to end-product inhibition of the enzymes. A cocktail of enzymes tailored to the feedstock and pretreatment method is essential. A pretreatment method that preserves much of the hemicellulose will require endo- and exo-xylanases and beta-xylosidase in the enzyme cocktail; conversely, a pretreatment that removes or solubilizes much of the hemicellulose (such as hot water extraction or solvent extraction) will need fewer (or different) enzymes targeted at hemicellulose hydrolysis. Accessory enzymes such as expansins and swollenins can also increase accessibility of hydrolytic enzymes and enhance conversion rates and yields. Most enzymes in the system are subject to inhibition by other carbohydrates, phenolics or organic acids generated during hydrolysis. Thus, strategies to limit inhibitor concentrations or remove them in situ are very important.

The commercial-scale producers of cellulosic ethanol have thus far made different choices regarding enzymes. Dupont is using enzymes produced via their own technology platform and Abengoa is using enzymes produced under license from Dyadic. Novozymes is in partnership with Beta Renewables to supply enzymes, and is also supplying enzymes to many of the other pilot- and commercial-scale producers of cellulosic ethanol.

Enzyme doses are highly dependent upon the type of substrate and the pretreatment method. Even with on-site enzyme production, processes that require an excessive enzyme dose are unlikely to be financially viable. Furthermore, a high enzyme dose can have a significant adverse effect on the GHG emissions profile of the cellulosic biofuel (Hong et al., 2013).

Fermentation

There has been much debate regarding the efficacy of different fermentation strategies for cellulosic ethanol production, and consequently, a variety of different fermentation organisms and process options have been evaluated and deployed at the commercial and pilot scale (Spatari et al., 2010; Humbird et al., 2011; Pourbafrani et al., 2014). Some processes are based upon separate trains for fermentation of C5 and C6 sugars; in this case, the C5 sugars are generated during pretreatment and sent to a C5 fermentation system, while the C6 sugars are generated during an enzymatic hydrolysis step prior to C6 fermentation. In this approach, a conventional yeast can be used to ferment the C6 sugars, while a novel/different fermentation organism is used to ferment the C5 sugars. In contrast, an alternative process design combines the C5 and C6 fermentation into a single set of fermenters, although the actual fermentation within the vessels may use the C6 and C5 sugars sequentially, rather than simultaneously. In such a design, the C5 and C6 sugars are generated during enzyme hydrolysis, and co-fed to the fermentation system.

A key design decision centers around the use of separate hydrolysis and fermentation steps (SHF), versus a combined process that is based upon simultaneous saccharification and cofermentation (SSCF). Much of the debate regarding these options centers upon the fact that most hydrolytic enzymes have temperature optima in the range of 50°C, whereas most fermentation organisms, particularly yeasts, have optima around 32–34°C. Unfortunately, the activity of many components in a "cellulase" cocktail drops off very quickly once the temperature falls below 45°C, which is a significant impediment to SSCF with yeasts at 32–34°C. On the other hand, many of these enzymes are subject to end-product inhibition, that is, the sugars that they produce inhibit the activity of the enzymes that produce them. Removing these sugars via a fermentation process can alleviate these inhibitory effects. However, the slow nature of the hydrolysis process is an impediment to SSCF, because insufficient sugars early in the process can limit cell growth and significantly reduce fermentation rates. Consequently, the most likely near-term option will be a hybrid—initiate hydrolysis under optimum "cellulase" conditions to produce sugars, then transfer to an SSCF process with enough sugars to launch the fermentation while also removing excess sugars that may be inhibitory.

Various attempts have also been made to increase the temperature tolerance of fermentation organisms, so that the SSCF process could take place at, for example, 40°C, where the enzymes are more active. However, this also dramatically increases the risk of contamination by, for example, lactobacilli and *Acetobacter* spp., which are much more active at 37 to 40°C than at 32°C.

Another key factor in the fermentation of cellulosic hydrolyzates is inhibitor tolerance. Almost all biomass contains acetyl groups that are released during pretreatment and hydrolysis, and the resulting acetate/acetic acid can dramatically reduce fermentation rates and yields. Management of acetate and other inhibitors is critical to the success of a fermentation process.

A variety of different fermentation organisms have been proposed for cellulosic ethanol production, based upon different capabilities to ferment xylose into ethanol and tolerance to inhibitors generated during pretreatment and/or released during hydrolysis. C5 fermenting capability has been added to yeast, while other options include modified *Escherichia coli*, Clostridia and *Zymomonas*. Irrespective of the choice, the fermentation organism must be able to deliver robust and consistent conversion of C5 and C6 sugars at high rates, yields and titers (which implies tolerance to ethanol as well).

In most cases, the fermentation is conducted in batch mode, using several fermenters operating in parallel. The presence of solids in the fermenters, especially during an SSCF process, can create significant challenges with mixing and the generation of inhibitors from phenolics. Unlike first-generation processes in which CO_2 generation during fermentation is sufficient to mix the fermenter contents, fermenters for second-generation ethanol will almost certainly include agitators or external pump-around systems to circulate the slurry. This is partly due to slower fermentation rates, but more importantly, due to the higher percentage of insoluble solids in the fermentation broth for second-generation processes.

Recovery and Use of Solid Residues

Similar to first-generation ethanol processes, solid residues are recovered from the bottom of the first distillation column. These residues are primarily comprised of lignin, but also include unconverted cellulose and hemicellulose, along with solids from cell matter. The solid residues include soluble solids such as unfermented carbohydrates and organic acids that are present in the fermentation broth and remain within the liquid fraction included with the recovered "solids." The solid residues are typically targeted as an energy source, but they are typically too wet to be used directly in a biomass combustion system. Thus, moisture is often removed via a combination of centrifugation, filtration and drying prior to delivery to a biomass boiler (Humbird et al., 2011).

A related challenge is the fact that biomass combustion and gasification systems are very capital-intensive, and, from a financial perspective, may not be the preferred method for processing biomass residues. Another option includes production of "lignin" pellets while producing process steam/thermal energy using a natural gas boiler. This lower capital cost option also allows the biofuel producer to capitalize on higher revenues from overseas wood pellet markets (Pourbafrani et al., 2014).

Coproduct Options: Technical, Financial and Market Effects

Different feedstocks and technology combinations can enable some high-value coproduct alternatives (Pourbafrani et al., 2014), rather than the typical focus on electricity as a coproduct. The lower temperatures used for alkaline pre-treatments such as ammonia–fiber expansion (AFEX) can preserve proteins and lignin, allowing these components to be purified and sold for food, feed or chemical applications. Pretreatments (or hydrolysis processes) that produce significant amounts of soluble hemicelluloses can be tailored to produce xylose or xylitol as coproducts. Generally, fewer coproduct options are available when acids are used as catalysts, due to additional degradation products created in the presence of acids. Furthermore, as noted above, pellets or electricity may be produced from the solid residues.

These alternate coproducts require additional capital and affect operating costs, typically by increasing process energy requirements. In each case, the potential added value of the coproduct must be weighed against the higher capital and operating costs. A secondary issue or opportunity can arise if the cellulosic fuel is covered under a low carbon fuel standard or GHG reduction regulations, for example, California's Low Carbon Fuel Standard. In this case, the magnitude of the GHG coproduct credit can vary significantly, depending upon the carbon intensity of the existing product displaced by the coproduct from the biorefinery. As shown by Pourbafrani et al. (2014), the GHG reduction on a well-to-wheels basis compared to gasoline can range between 60 and 140%, depending upon the process technology, feedstock and coproduct.

Performance and Life Cycle Financial and GHG Emissions Impacts

Three key metrics dictate the financial feasibility of second-generation technologies: (1) ethanol titers, (2) ethanol yield and (3) production rate. High yields indicate efficient use of the biomass feedstock, whereas high production rates tend to reduce capital costs, and high titers are indicative of high solid loadings that will reduce the hydraulic load (and capital cost), while also reducing process energy demand. It is rare to maximize all three metrics, but some tradeoffs can strike a cost-effective balance. Based upon reported production capacities and feedstock requirements, current commercial technologies are reporting yields in the range of 70–85 USG/dry tonne of feedstock, and fermentation titers in the range of 6–8 wt%. It is likely that both metrics will improve as the industry matures.

As noted above, predicted GHG emission reductions, compared to gasoline, can range between 60 and 140%, but these values are based upon theoretical processes/models, and may not accurately reflect the tradeoffs between rates, titers and yields that are part of a commercial plant design. Very few published data are available to document GHG reductions for commercial-scale plants.

However, the GranBio application under California's LCFS indicates a carbon intensity of $7\,gCO_{2eq}$/MJ for production of ethanol from sugarcane straw, with an electricity credit of 2.98 kWh/gal (California Air Resources Board, 2014). Similarly, the LCFS application for Abengoa's Hugoton biorefinery cites emissions of $23.4\,gCO_{2eq}$/MJ for wheat straw and $29.5\,gCO_{2eq}$/MJ for corn stover (California Air Resources Board, 2015).

Additional factors that can influence the GHG intensity of second-generation ethanol production include the type of feedstock, the ethanol yield, and the location of the plant, particularly for biorefineries that also produce electricity. The effects of feedstock and plant location are related, because the feedstock yield and the intensity of the farming operations to produce and harvest the feedstock vary geographically, and will influence the GHG emissions attributed to the feedstock. Furthermore, the carbon intensity of the grid varies by geography, affecting the magnitude of the GHG credit for the electricity coproduct.

Although it seems counterintuitive, a lower yield leads to a better GHG performance for the cellulosic ethanol, because additional electricity is produced. Since the carbon intensity of the grid electricity in many jurisdictions is often greater than that of gasoline, a process that produces more electricity and less ethanol can have a better GHG profile for its ethanol product, due to the coproduct credit generated by the renewable electricity (Pourbafrani et al., 2014; Shen, 2012).

SYNOPSIS OF COMMERCIAL PLANTS

A small number of commercial plants have been built or are under construction, with some other technology providers marketing commercial plant designs based upon pilot-scale operations.

The first modern, commercial-scale cellulosic ethanol plant was built by Beta Renewables in Crescentino, Italy. The plant is designed to use either wheat straw or *Arundo donax* as feedstocks, and has a reported capacity of 40,000 or 60,000 tonnes/year of ethanol (depending upon feedstock). The plant, which also produces renewable electricity, has been operating since 2012. The process is based upon autohydrolysis pretreatment, and uses enzymes from Novozymes, coinciding with Novozymes' investment in Beta Renewables.

GranBio licensed the Beta Renewables technology to produce second-generation ethanol from sugarcane straw in Brazil (Beta Renewables GranBio Project, 2015). This plant, designed to produce 65,000 tonnes of ethanol annually, is reported to be in the commissioning phase (GranBio, 2014). GranBio has also developed their own proprietary yeast for second-generation ethanol production, to satisfy the stringent regulatory requirements in Brazil (GranBio, 2015).

Soon thereafter, Dupont, Abengoa and POET-DSM announced plans to construct plants using their respective technologies. All are based upon use of crop residues—either wheat straw or corn stover. There are notable differences between the technologies. Dupont relies on an alkaline pretreatment, and uses "cellulase" enzymes developed internally by their enzyme subdivision. Abengoa's process uses a dilute acid pretreatment in conjunction with steam explosion, and uses enzymes licensed from Dyadic. The targeted plant capacity is 25 million gallons (95 million liters)/year from 1000 tons (907 tonnes)/day of biomass, with 21 MW of electricity coproduct (Abengoa, 2014).

POET-DSM is also using dilute acid pretreatment combined with steam explosion. Long term, DSM is supposed to provide enzymes and fermentation organisms for the second-generation process. The POET-DSM facility is colocated with an existing POET corn-ethanol plant, which allows some sharing of infrastructure and utilities, to the benefit of both plants. The POET-DSM facility, which opened in September, 2014, and as of this writing, is continuing to ramp up operations, aims to convert 770 tons (700 tonnes)/day of corn stover into 20 million gallons (76 million liters)/year of ethanol (DSM, 2014).

Raizen/Iogen are codeveloping a 40 million liter per year (MMLY) facility, located adjacent to Raizen's existing Costa Pinto sugarcane mill in Brazil, based upon the R9 generation Iogen technology developed and validated at Iogen's pilot facility in Ottawa, Canada (Costa Pinto Project). The process also uses dilute acid pretreatment in combination with steam explosion to process sugarcane bagasse. Raizen/Iogen reported a successful start-up of the facility in December 2014 (Raizen, 2014).

FUTURE DEVELOPMENTS FOR CELLULOSIC ETHANOL

Increasing Scale: Technology and Efficiency Improvements

Most of the current second-generation technologies use dilute acid pretreatment in combination with steam explosion. The use of dilute acid is driven by the need to enhance hydrolysis efficiency and compensate for the challenges still encountered with enzyme use—primarily high doses or slow rates, coupled with inhibition by compounds generated during pretreatment and hydrolysis. As enzyme and pretreatment technologies improve, it is likely that dilute acid will be phased out in favor of autohydrolysis or alkaline pretreatments that have much lower capital and operating costs. These improvements will also facilitate use of higher solid loadings during hydrolysis, with concomitant reductions in equipment size and higher product titers that reduce energy use.

A second likely area of improvement relates to fermentation, both in terms of C5 utilization and tolerance to inhibitors. More efficient and complete utilization of xylose and other C5 sugars by fermentation organisms will increase yields and fermentation rates, reducing capital and operating costs while also reducing the risk of infection associated with slow fermentations. Similar improvements are also likely to arise by engineering microbial tolerance to organic acids and phenolics released during pretreatment and hydrolysis.

Capital costs for current commercial-scale plants have been reported to be in excess of $8/installed gallon of capacity, about fourfold greater than the cost for a corn ethanol plant. Some of these differences are inevitable because of the different types of feedstocks, but improvements in process efficiency via better enzymes and fermentation organisms can significantly close the gap.

Consider that corn is 70% starch, over 90% of which is efficiently converted to sugars, and then to ethanol. This means that the equipment must be sized to also carry the ~30% of the feedstock that is unconverted (unless removed upfront by fractionation) and eventually ends up as a coproduct. Ethanol titers in the range of 16–20 wt% are common within 50 h of fermentation, implying smaller fermenters and leading to efficient distillation with a lower energy demand.

By comparison, lignocellulosic biomass is often 55–65% carbohydrate and 35–45% lignin. Current processes convert about 70% of the available carbohydrate into fuel. Thus, there is a much higher percentage of noncarbohydrate components that must be carried through the process, along with a much higher percentage of unconverted carbohydrate. This adds substantially to the size of the equipment, and increases the energy needed to pump/transfer material between process units. The higher hydraulic loads and lower carbohydrate conversion mean lower ethanol titers, typically 5–8 wt%, coupled with slower fermentations. Improvements to process technology, enzymes and fermentation organisms that increase yields, hydrolysis and fermentation rates, and solids loadings are key to closing the current gap in capital and operating costs between lignocellulosic ethanol and corn ethanol plants.

Role of Feedstocks in Industry Expansion

The initial lignocellulosic ethanol plants are using crop residues, particularly corn stover, wheat straw and sugarcane bagasse as feedstock. While useful, the field density of these residues is comparatively low, and the need for long-term storage following a short harvest season poses logistical challenges. As the industry grows, it is likely that it will expand its use of dedicated energy feedstocks that have a higher yield and longer harvest window. That said, these novel feedstocks need to be derisked for both growers and end-users. For example, it is important to consider the need for crop insurance, the risks and benefits of a multiyear crop such as energy cane versus an annual crop such as sorghum, and the ease (or lack thereof) in scaling up crop production as new plants come online, and matching the timing of feedstock supply with the timing of the start-up of the biofuel facility. These are onerous challenges while the industry is in its infancy, but will diminish as new plants come online.

Colocation and Other Synergies

There are many opportunities to capitalize on synergies with existing or colocated facilities. As noted above, the POET-DSM facility can share infrastructure and utilities with the existing corn-ethanol plant. For example, if there is excess electricity or steam from the lignin generated during cellulosic ethanol production, these utilities could be used at the nearby corn ethanol plant, thereby reducing its carbon footprint and energy cost. Colocation of a cellulosic ethanol facility with a pulp mill can provide access to wood residues and utilities, while sharing storage and transportation infrastructure and logistics (McKechnie et al., 2011).

In Brazil, second-generation ethanol facilities may be colocated beside existing sugar mills, utilizing the bagasse residues, but also capitalizing on the potential to use fermentation and distillation equipment that might otherwise sit idle during the cane "off-season." This is a key attribute of the Raizen/ Iogen facility.

There can also be synergistic opportunities on the coproduct side. For example, the cellulosic ethanol plant could be colocated with a pellet mill that processes excess lignin residues into "pellets" that could be blended with (or replace) coal. The lignocellulosic ethanol plants also, as an intermediate step, produce a variety of sugars, some of which could be isolated and sold as higher-value products, for example, to produce polyols. Many of the opportunities are in small markets, and thus, recognition of local market conditions and careful evaluation are required.

FINAL REMARKS

The North American ethanol market is poised for a transition to second-generation technology, building upon its longstanding growth and success in first-generation processes that convert feed corn into ethanol. The pioneering new plants coming online capitalize on crop residues, providing synergies with existing field crops such as corn and wheat. Eventually, dedicated energy crops such as switchgrass, energy sorghum and energy cane are expected to play a role because of their enhanced productivity, lower input requirements and expanded harvest seasons. These new feedstocks, coupled with enhancements to process design and efficiency and strategic development of coproducts, can form the foundation for an era of low-carbon fuels and high-value bioproducts produced from renewable resources.

REFERENCES

Abengoa, 2014. Abengoa celebrates grand opening of its first commercial-scale next generation bio-fuels plant. Press Release. http://www.abengoa.com/web/en/noticias_y_publicaciones/noticias/historico/2014/10_octubre/abg_20141017.html.

Beta Renewables GranBio Project, 2015. http://www.betarenewables.com/projects/5/granbio (accessed 22.09.15.).

Brummels, T., Saville, B.A., 2015. Attributes of energy cane for biofuel and bioproduct production. In: Presented at the Advanced Biofuels Feedstock Conference, New Orleans.

California Air Resources Board, 2014. http://www.arb.ca.gov/fuels/lcfs/2a2b/apps/gb-rpt-082514.pdf (accessed October 2015).

California Air Resources Board, 2015. http://www.arb.ca.gov/fuels/lcfs/2a2b/061615lcfs_apps_sum.pdf (accessed October 2015).

Chandra, R., Bura, R., Mabee, W., Berlin, A., Pan, X., Saddler, J., 2007. Substrate pretreatment: the key to effective enzymatic hydrolysis of lignocellulosics? Adv. Biochem Engin./Biotechnol. 108, 67–93.

Costa Pinto Project. http://www.iogen.ca/raizen-project/ (accessed 07.10.15.).

Day, D., 2015. New crops for biofuel/bioproduct production and the feedstock readiness profile. In: Presented at the Advanced Biofuels Feedstock Conference, New Orleans.

Di Risio, S., Hu, C.S., Saville, B.A., Liao, D., Lortie, J., 2011. Large sacle. High-solids enzymatic hydrolysis of steam-exploded poplar. Biofuel. Bioprod. Bior. 5, 609–620.

DSM, 2014. First Commercial Scale Cellulosic Ethanol Plant in the US Opens for Business. Press Release. http://www.dsm.com/corporate/media/informationcenter-news/2014/09/29-14-first-commercial-scale-cellulosic-ethanol-plant-in-the-united-states-open-for-business.html.

Ethanol Producer Magazine, 2015. BBI. http://www.ethanolproducer.com/plants/listplants/US/Existing/Sugar-Starch/.

Galbe, M., Zacchi, G., 2007. Pretreatmentof lignocellulosic materials for efficient bioethanol production. Adv. Biochem. Engin./Biotechnol. 108, 41–65.

Gill, J., Burks, P., Staggenborg, S., Odvody, G., Heiniger, R., Macoon, B., et al., 2014. Yeild results and stability analysis from the sorghum regional biomass feedstock trial. Bioenergy Res. 7 (3), 1026–1034.

GranBio, 2014. GranBio Begins Producing Second Generation Ethanol. Press Release. http://www.granbio.com.br/en/blog/granbio-begins-producing-second-generation-ethanol/ (accessed 22.09.15.).

GranBio, 2015. GranBio Obtains Commercial Approval for Its First Proprietary Yeast. Press Release. http://www.granbio.com.br/en/blog/granbio-obtains-commercial-approval-for-its-first-proprietary-yeast-2/ (accessed 22.09.15.).

Greenshift. Corn Oil Extraction. http://www.greenshift.com/products/corn_oil_extraction (accessed 10.10.15.).

Hong, Y., Nizami, A., Bafrani, M.P., Saville, B.A., MacLean, H.L., 2013. Impact of cellulase production on environmental and financial metrics for lignocellulosic ethanol. Biofuel. Bioprod. Bior. 7, 303–313.

Hu, G., Heitmann, J., Rojas, O., 2008. Feedstock pretreatment strategies for producing ethanol from wood, bark and forest residues. Bioresources 3 (1), 270–294.

Huang, H., Liu, W., Singh, V., Danao, M.-G., Eckhoff, S., 2012. Cereal Chem. 89 (4), 217–221.

Humbird D, Davis R, Tao L., Kinchin C., Hsu D., Aden A., et al., 2011. Process Design and Economics for Biochemical Conversion of Lignocellulosic Biomass to Ethanol.

ICM Inc., Fiber Separation Technology (FSTTM). http://www.icminc.com/products/fiber-separation-technology.html, https://www.youtube.com/watch?v=yjkIfLFY81o (accessed 01.10.15.).

Jessen, H., 2014. Ethanol Producer Magazine, BBI, Beyond Corn Oil Extraction. www.grains.org/sites/default/files/ddgs-handbook/Chapter-2.pdf.

Kim, S., Dale, B., 2015. Comparing alternative cellulosic biomass refining systems: centralized versus distributed processing systems. Biomass Bioenergy 74, 135–147.

McKechnie, J., Zhang, Y., Ogino, A., Saville, B., Sleep, S., Turner, M., et al., 2011. Impacts of co-location, co-production and process energy source on life cycle energy use and greenhouse gas emissions of cellulosic ethanol. Biofuel. Bioprod. Bior. (BioFPR) 5, 279–292.

Mosier, N., Wyman, C., Dale, B., et al., 2005. Features of promising technologies for pretreatment of lignocellulosic biomass. Bioresour. Technol. 96, 673–686.

POET. http://www.poet.com/advantage (accessed 10.09.15.).

POET-DSM. Responsible Corn Stover Harvest. http://poet-dsm.com/resources/docs/responsible-stover-harvest.pdf (accessed 07.10.15.).

Pourbafrani, M., McKechnie, J., Shen, T., Saville, B.A., Maclean, H.L., 2014. Impacts of pre-treatment technologies and co-products on greenhouse gas emissions and energy use of lignocellulosic ethanol production. J. Clean. Prod. 78, 104–111.

Raizen, 2014. Iogen Commence Cellulosic Ethanol Production in Brazil, Biofuels Digest. http://www.biofuelsdigest.com/bdigest/2014/12/17/raizen-iogen-commence-cellulosic-ethanol-production-in-brazil/.

Sanchez, O., Cardona, C.A., 2008. Trends in biotechnological production of fuel ethanol from different feedstocks. Bioresour. Technol. 99, 5270–5295.

Saville, B.A., 2011. Pretreatment options Plant Biomass Conversion. John Wiley & Sons, Inc.

Shen, T., 2012. Life Cycle Modelling of Multi-Product Lignocellulosic Ethanol Systems (M.A.Sc. thesis). Department of Chemical Engineering and Applied Chemistry, University of Toronto, Toronto.

Spatari, S., Bagley, D.M., MacLean, H.L., 2010. Life cycle evaluation of emerging lignocellulosic ethanol conversion technologies. Bioresour. Technol. 101, 654–667.

Sun, Y., Cheng, J., 2002. Hydrolysis of lignocellulosic materials for ethanol production: a review. Bioresour. Technol. 83, 1–11.

US Grains Council, Chapter 2, Ethanol Production and Its Co-Products, www.grains.org/sites/default/files/ddgs-handbook/Chapter-2.pdf (accessed 01.10.15.).

US Renewable Fuels Association (RFA), 2015. http://www.ethanolrfa.org/resources/biorefinery-locations/.

Wullschleger, S.D., Davis, E.B., Borsuk, M.E., Gunderson, C.A., Lynd, L.R., 2010. Biomass production in switchgrass across the United States: database description and determinants of yield. Agron. J. 102 (4), 1158–1168.

Chapter 8

Technological Foresight of the Bioethanol Case

José Maria F.J. da Silveira[1], M.E.S. Dal Poz[2], L.G. Antonio de Souza[3] and I.R.L. Huamani[1]
[1]Institute of Economics, State University of Campinas, Campinas, São Paulo, Brazil
[2]Faculty of Applied Sciences, State University of Campinas, Limeira, São Paulo, Brazil
[3]Interdisciplinary Center of Energy Planning, State University of Campinas, Campinas, São Paulo, Brazil

INTRODUCTION

Energy research fields have been defined on the basis of a systematic application of the technical and scientific knowledge of agents who compete in a context of selective market processes, characterizing a science based area of economy (Bell and Pavitt, 1993). These processes take different trajectories and generate different patterns of technology diffusion within firms, between firms and among firms, in the same industry, between industries and sectors in different countries, particularly the USA and Brazil, the leading countries in ethanol production nowadays.

The tight connection between scientific and technological development and innovation justifies the importance of mapping the emerging areas of research of bioenergy (Brown and Brown, 2012; Souza et al., 2015); the identification of technological trajectories that can point to convergence or divergence of the technological alternatives and its level of maturity (Fontana et al., 2009; Dal Poz et al., 2013; Dal Poz and Silveira, 2015; Ferrari, 2015) and finally, find evidences of specialization in complex systems (Frenken, 2006; Murakami et al., 2015).

The central idea of the paper is to build networks of scientific publication and patents, and moreover, to show some empirical results of those methodologies to fields related to bioenergy, particularly, biofuels. These procedures involve the design and application of two groups of foresight tools: (1) Global scientific and technological production and networks of scientific collaboration, using scientometric and bibliometric approaches for "second-generation bioethanol

R&D global efforts"; and (2) Networks of forward patent citations as proxies of innovation potential, concerning one example of a huge technological demand: biomass raw material industrial processes for "ethanol or bioethanol."

The combinations of foresight methodologies are open to identify signal and noises that point to long-term co-existence of technological alternatives for biofuel production, raising uncertainty. Conversely, it also turns to the identification of trends and of technological convergence to guide investments and innovation.

The foresight application of cases concerning "bioethanol R&D" are demonstrations of methodological tools and the possibilities to integrate them. However, these investigations have potential to be the base for a link between different kinds of elements in the bioethanol scene: (1) scientists; (2) institutions; (3) technologies; (4) raw materials; and (5) firms. The main dimensions of these connections are: (1) scientific collaboration; (2) mechanisms of appropriability (patents, secrecy); (3) technology transfer agreements; (4) *brown* and *greenfield* investments; and (5) upstream investments.

Section "Collecting Information From Data Banks" also presents a brief methodological introduction involving procedures that are required in any empirical exercise.

In Sections "Applying Scientometrics to Scientific Papers in the Bioenergy Sector: Network Analysis in Second-Generation Ethanol" and "Using Patents to Forecast Technological Trends," two cases of the technological foresight methodological applications for bioethanol will be presented: (1) ethanol/bioethanol publication data to build scientific collaborative networks; and (2) technological trajectories for bioethanol based on patent data, comprising two studies: (1) using network analysis and (2) by the application of a specialization index to identify co-occurrence of three selected dimensions related to enzymatic hydrolysis, one of the most promising technologies for second-generation ethanol.

Conclusions and policy recommendations for international institutions and for the selected "national systems of innovation" with interest on biofuels will be included in the final section of the chapter.

COLLECTING INFORMATION FROM DATA BANKS

Souza et al. (2015) and Souza (2013) point to the following procedures to build a data bank for bibliometric and scientometric research using scientific publications:

1. Collect data on publications and citations Web of Science webset (WoS),[1] which keeps databanks of citations that include thousands of academic journals and offer bibliographical database services;

---

1. WOS is in a site from Thomson Reuters.

2. The crucial step is to define the structure of the queries, using *boolean* logic. An example of two complex queries respectively by Souza et al. (2015) and Souza et al. (2014), to retrieve data from WoS in bioenergy field.[2] In this case, the task is to combine words selected from the state-of-art literature in lignocellulosic ethanol, and those words expected to be present in many scientific articles, like energy, and specific words like enzymatic hydrolysis;

3. Use a program that can translate data previously extracted in a format (example, format ".txt" with no quotes) and then enable analysis or even export data filtered (Souza et al., 2015). The VantagePoint[3] program makes it possible to import the information obtained from the WoS/ISI through filters;

4. After importing data, precleaning took place aiming to extract information and even duplicate the grouping of terms that for some reason the original base had wrongly drafted.

To build a data bank of patents few, but important, changes are necessary to be implemented. Firstly, the source is now a patent data bank, like USPTO or Derwent, among others.[4] Secondly, the choice of the sections of the patent to perform the search procedure is critical: in the title, abstract, claims, each one separately or all items together. The greater is the number of sections the researcher wants to include, the better is patent coverage, the more costly is the procedure in terms of time to complete the task.

The use of filters is a strategy to shape the databank, but needs care. The easiest way to filter documents (scientific papers or patents) is to censor the words that compose the queries. Dal Poz et al. (2013) aiming to identify patents related to genetically modified crops, limited the keywords to the name of promoters, like *35S* and/or *Ubiquitin* instead of using, for instance, *transgenic crops*. Doing this the procedure avoids documents dealing with socioeconomic aspects or patents of cultivars: the focus is put on science.[5]

Retrieving a large number of patents makes the analysis troublesome, demanding appropriate strategies to deal with this. There are many possibilities to reduce the size of the data bank after the use of the search procedures. For instance, building "clusters" from a giant component of a network allows the identification of homogeneous groups and the evaluation of their stability in a

---

2. Having skills and knowledge of the object are the conditions to sum up in a few words the content of the full text. The aim is to capture the best lexical-semantic relations from a full-text database (Vorhees, 1994).

3. The *VantagePoint* version 7, http://www.thevantagepoint.com/ has been used in the paper.

4. There are more comprehensive data banks, resulting from the merger of data from different patent offices, like the *Pat Base*. Although they are more complete in terms of coverage of applicants, assignees and places where the patents had been filed, there are some difficulties to collecting data from them using "robot", due to the format of the texts. See Masago (2012).

5. Dal Poz et al. (2013) have classified the keywords in layers, some of them placed in the macro level, like energy, others in mesolevel, as promoters and others placed at microlevel, like "simultaneous saccharification."

timeline (Shibata et al., 2008). The use of the International Patent Classification is critical to gather a selected group of patents to apply an entropy index of specialization (Murakami et al., 2015).

Summing up, filters are useful but require some degree of intimacy with the subject of analysis. An exploratory study demands a good definition of keywords and careful measures to shape the data bank.

The next sections will present the results of one application of bibliometrics and scientometrics to the bioenergy fields.

APPLYING SCIENTOMETRICS TO SCIENTIFIC PAPERS IN THE BIOENERGY SECTOR: NETWORK ANALYSIS IN SECOND-GENERATION ETHANOL[6]

The conception of intellectual connections between the ideas of scientists is established through social relations, and, in certain areas, the collaboration is crucial to development of a specific research field. There is robust evidence showing that the scientific collaboration—authors of coauthored publications and citations that reference other authors in their publications—are positively correlated with the diffusion of scientific knowledge.

The main objective is to describe the international network of scientific papers in bioenergy, looking for its remarkable characteristics, and establishing a discussion on trends and policies. It is possible to argue that the use of network analysis made possible a broader view of the bioenergy scene, in comparison with the description of the publications only using qualitative tools. However, in this section, the methodology applied to a more restricted field is presented, strongly based in the work of Souza et al. (2015), presenting the results of a procedure to an investigation on second-generation ethanol.

The group of keywords to build the search parameter combines words selected from the state-of-the-art literature in lignocellulosic ethanol, energy with specific words like enzymatic hydrolysis. After adopting an analysis of the words from the perspective of three panels of experts in bioenergy, particularly in bioethanol, a query is composed with terms of the investigation. The query is presented in Annex I.

Finally, the network approach allows the use of indicators in two basic levels: (1) the whole network and (2) the individual level.

The following relationships in collaborative networks (CNs) are presented in Figs. 8.1 and 8.2:[7] the relationship between countries (macrolevel)

---

6. The bibliometrics and scientometrics studies have specifications, approaches and different roles among themselves. The objective of bibliometrics is to study books or journals, in order to understand the activities of science of information. The aim of scientometrics is to study the quantitative aspects of the creation, dissemination and use of scientific and technical information and the objective is based on understanding search engines such as social activity.
7. Following Souza et al. (2015), the Fruchterman–Reingold algorithm was chosen for the display of the networks, as it best represented the data because of the enormous quantity of relationships.

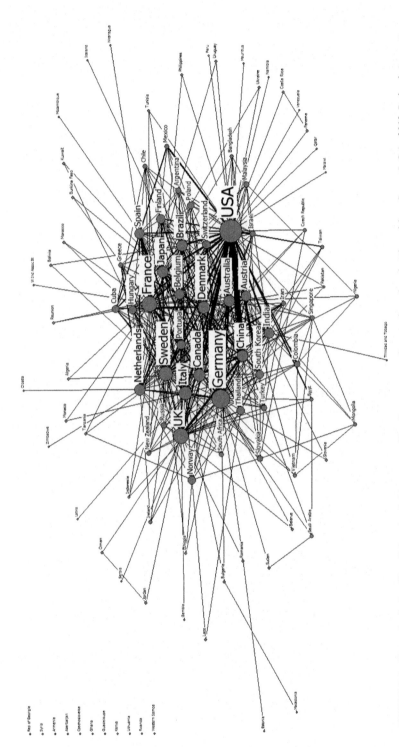

FIGURE 8.1 Network of scientific collaboration in bioenergy, with a focus on second-generation ethanol. *Adapted from Souza, L.G.A., 2013. Redes de inovação em etanol de segunda geração. Tese (doutorado) -Universidade de São Paulo, Escola Superior de Agricultura "Luiz de Queiroz". Piracicaba, São Paulo, using VantagePoint and UCINET.*

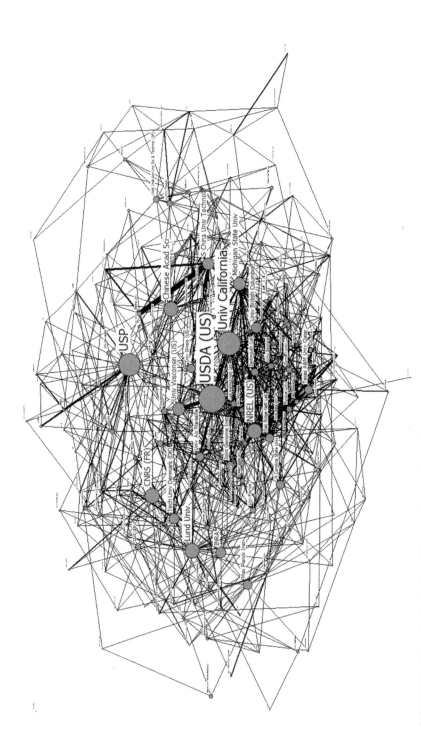

FIGURE 8.2 Network of scientific publications in second-generation ethanol by organization. *Adapted from Souza, L.G.A., 2013. Redes de inovação em etanol de segunda geração. Tese (doutorado) - Universidade de São Paulo, Escola Superior de Agricultura "Luiz de Queiroz". Piracicaba, São Paulo, using VantagePoint and UCINET.*

and partnerships between institutions, that is, universities, governments and firms (microlevel). They have taken part directly or indirectly in the knowledge production of second-generation ethanol. A total of 6948 papers have been retrieved, with 2286 in collaboration between countries, representing 32.90%. Exactly 103 countries have been included in the giant component of the network, presented in Fig. 8.1 and few isolates (less 10) has been found.

Table 8.1 presents the results of the top 10 countries in number of scientific publications and share of papers in international collaboration: the relevance of collaboration for the country and the share of collaboration in the total of papers retrieved by Souza (2013). A total amount (for 10 countries) of 4501 papers were retrieved, with 1285 as result of an international collaboration among then. The 10 top countries are responsible for 65% of the papers published.

As expected, EUA has published up to 35% of the papers in this field, followed by China and Brazil, respectively with 15% and 7.7%. The gap between Brazil and the others has reduced sharply. Germany, Canada, the UK and Sweden have more than 40% of their publications in collaboration, but probably not with themselves, showing a center–periphery pattern.

Taking a look in the last column of Table 8.1, it is possible to see that although collaboration has less importance to the United States and China (less than 30% of the papers), they have the first two positions in the rank of collaboration of the world, followed by Germany, Sweden and the UK.

The importance of the top 10 countries presented in Table 8.1 is clear in Fig. 8.1: they are placed in the center of the network, meaning that they are not only involved in the majority of collaborations, but that they are the countries that connect other countries.

Wagner and Leydesdorff (2005) analyzed the scientific publications structured in international knowledge networks—in which authorship and coauthorship play the role of links among actors. According to them, the behavior of networks of collaboration can be described by the behavior of the preferential attachment, which is based on the reputation and rewards. This gives a good explanation of the role of the countries with the higher indexes of centrality in the network: they are able to connect to some countries and exclude others, and just because of this, make their partners (geodesic distance 1) attractive to other countries.

The size of the node (vertex) in Fig. 8.1 represents the degree of each country[8] and the thickness of the edge the strength of collaboration between countries. There is a clear triad between EUA, China and Germany, the first two contributing with the largest knowledge cumulated in the area and the last, as an adjunct partner. In a similar way, EUA, China and the UK have an important

---

8. The total degree of a country is the number of connections the country has with other countries. For instance, France is not included in the top 10, but has a higher degree than other countries in Table 8.1.

TABLE 8.1 Number of Papers, Number of Papers in International Collaboration, and Its Relative Importance to the Country and to the Total Number of Mapers Published in Second Generation (From the Start to 2012)

Country	Publications (Number)	Total (%)	National	(%)	Collaboration (Number)	(%)	Collaboration (Relative Importance) (%)
USA	1,559	22.44	1,190	76.33	369	23.67	16.14
China	684	9.84	505	73.83	179	26.17	7.83
Brazil	**347**	**4.99**	**257**	**74.06**	**90**	**25.94**	**3.94**
Japan	332	4.78	249	75.00	83	25.00	3.63
India	299	4.30	250	83.61	49	16.39	2.14
Germany	290	4.17	169	58.28	121	41.72	5.29
Canada	275	3.96	178	64.73	97	35.27	4.24
UK	244	3.51	133	54.51	111	45.49	4.86
Spain	240	3.45	169	70.42	71	29.58	3.11
Sweden	231	3.32	116	50.22	115	49.78	5.03

Source: from Souza, L.G.A., 2013. Redes de inovação em etanol de segunda geração. Tese (doutorado) - Universidade de São Paulo, Escola Superior de Agricultura "Luiz de Queiroz". Piracicaba, São Paulo.

connection. Fig. 8.1 shows triads between EUA, Germany and Brazil; and EUA, Finland and Germany. Showing the importance of network analysis, Germany and UK are respectively in the sixth and eighth places in number of publications, but in second and third places in a rank based on the index of normalized centrality.

These triads reflect a mix of scientific collaboration and technological interest in the field of second-generation ethanol, particularly to explore natural endowments, like wood, bagasse or corn stove. There are also regional networks, like EUA, Mexico and Cuba, but they weaker in comparison to the main network.

The partnerships among institutions (universities, government and enterprises— international collaboration networks) have resulted in a network in a microlevel, whose vertex are linked by edges of coauthorships.

Fig. 8.2 presents the international collaborative network of coauthorship between institutions in the research field of second-generation ethanol. There are many different types of organizations, universities, public laboratories and private companies. As expected, the more prominent institutions in Fig. 8.2 were from the main countries shown in Fig. 8.1 (like the USA—USDA, NREL and University of California). The central position of USDA confirms the relevance of the network approach: national energy acts combined with incentives to bioenergy research are the main feature of US policy and its proximity with University of California, research institutes like National Renewable Energy Laboratory (NREL), with the Chinese Academy of Sciences and the private company Novozymes reveals a strong attractor. There are regional subnetworks revealed, for instance, by the strong connection between Chinese Academy of Sciences and University of Beijing; or the "France–Italy" connection between CNRS, INRA, University of Lyon and University of Marseille and University of Genova.

In fact, organizations working with a broader range of subjects were linked in particular ways with specialized labs, placed in countries with specialized knowledge and skills. As pointed out by Bueno et al. (2015), Brazil looks better connected when the network highlights public institutions.

The University of São Paulo (USP) is an important hub in bioenergy (ranks second in terms of the index of centrality), connecting Brazilian institutions (public in the majority) with international research centers.

However, Souza et al. (2015) pessimistically concluded that the other Brazilian organizations are out of the "top 10" rank, characterizing a local subnetwork that reinforces the importance of USP. A local subnetwork is clear by the strength of the connections between USP, UNESP and Unicamp, the three most important universities in São Paulo State, and top 5 in the rank of universities of Brazil. Internationally, USP is strongly linked to the University of California and USDA, denoting prestige. USP, Unicamp and TU-Delft appear as an interesting triad, characterizing a "small world" condition: when local subnetworks look for knowledge in a distant player.

The richness of the network analysis becomes clear with the possibility to unfold the approach to other dimensions and keep the links amongst them. Souza (2013) and Souza et al. (2015) have built the collaborative network for "KeyWords Plus" based on scientific publications, with the expected results: the highlighted words are ethanol, ethanol production, biomass and fermentation (maybe, the unique exception). With less emphasis (but with greater importance *Saccharomyces cerevisiae* (fungus), hydrolysis, cellulose, wheat straw, *Escherichia coli* (bacteria), corn straw and pretreatment. The authors conclude that fermentation is a highlighted word due to the importance to improve fermentation in second-generation processes (using xylose, for example).

Sometimes, the absence of an item—*Saccharum* and/or sugar cane—gives a clue for analysis, ratifying the idea that Brazilian collaboration is limited. Bueno et al. (2015) show that Brazil is the most influential country when the query includes these words, with connections with USA, India, South Africa and Australia, a subnetwork of the network presented in Fig. 8.1 This raises the question of the complementarity of knowledge in the field of second-generation research. For instance, Dal Poz et al. (2013) have called attention to the knowledge transfer from USA to Brazil in second-generation ethanol.

USING PATENTS TO FORECAST TECHNOLOGICAL TRENDS

This section has the basic assumption that patent citations can be considered networks of innovation, since there is extensive empirical evidence demonstrating that a high degree of citation for a patent is correlated with market presence, which allows us to consider highly cited patents as examples of innovation (Hall et al., 2001; Jaffe and Trajtemberg, 2002). The main hypothesis for the use of this method is that frequent citations indicate that a patent contains relevant information for future inventions, thus, it is considered as technologically important. Therefore, the analysis of the structure of the patent citation network allows the identification of central patents as well as main trajectories that characterize the evolution of a specific technological field. However, as many experts have pointed out in scientific meetings, a good result, allowing a consistent forecast, depends on the fact that technological trajectories have already been on course. When the activities of patenting are at their very beginnings, it would be wrong to find any consolidated path.

The methodological framework to calculate search path link count (SPLC), pioneered by Hummon and Dorein (1989) and the applications to economics, like the work of Verspagen (2007) for fuel cells and Fontana et al. (2009) for communication systems, are the main references of the exercise to illustrate the role of networks of patents in the identification of technological trajectories (TT) in bioenergy.

TT is identified using SPLC index (Fontana et al., 2009), which assigns to each link a value that quantifies the number of different paths passing through the arc, assuming that frequent relations are strategic for technological development, and that patents involved contain relevant information for future inventions.[9] Thus, they are considered as technologically important in the standpoint of "main flows of knowledge" that characterizes the field of bioenergy. In other words, the link of patents with higher SPLC is a kind a "valley" that makes easier the link between patents that can be considered original of a trajectory and those that represent the emergent technologies (Souza et al., 2014).

The calculation of the scores of different trajectories has as a starting point a patent citation network on bioenergy, retrieved from USPTO data bank, from 1976 to 2012 and using the software Odissey.[10] The keywords used were bioethanol/ethanol, with no filters, a macro perspective according to Dal Poz et al. (2013).[11]

An inspection into the contents of patents shows the existence of a giant component, as expected, related to bioenergy or energy. Analyzing only the giant component a total of 2018 patents was verified (1006 patents has been removed, those that were not related with the giant component), with 8673 links and 221,991 paths. The giant component was carefully inspected to check the pertinence of the content of patents to the subject of the investigation.[12]

The general indicators show figures similar to those related to the previous networks presented in Section "Collecting Information From Data Banks." The average distance among reachable pairs (average path length (APL)[13]) is 2.88, pointing to an easier contact between vertices (patents), which means better channels of information, what is expected according to the literature. The dimension of the giant component is nine meaning that every patent needs at most nine patents reach any other patent of the giant component (Jackson, 2010).

An expressive number of 514,572 TT was obtained using SPLC. The selection process was based on the weight of components of a trajectory. The results show that 90% of the patents in the giant component present an in-degree from 0 up to 13 but only one patent presented the highest one. SPLC values vary from zero to 0.1354, but only three citations present an SPLC exceeding 0.1053.

---

9. See Nooy et al. (2005) for the details of the calculation of SPLC and APL.
10. Odissey is a software developed by Fabio Masago dedicated to retrieving patents from databank (USPTO) and building citation networks composed by vertices and arcs. See Masago (2012). Nowadays VantagePoint is also able to perform similar tasks.
11. Although the absence of filters has had the goal of obtaining a broader network related to the subject, a second step has been done to avoid an overcharge in the system: the isolation and inspection of a giant component, in the assumption that it corresponds to the use of ethanol in energy. The procedure was based on the identification of subnetworks that were star-shaped, with some vertices with very high value of in-betweenness and of the centrality indexes. At least six or seven subnetworks have been identified, linked to the giant component by bridges (vertices), to be removed. The content of those sub-networks were: (1) ethanol for medical procedures; (2) solvent; (3) detergent; (4) diagnosis; (5) food industry; and (6) health and personal care.
12. Giant component means that all of the 2018 patents are connected.
13. The average number of steps along the shortest paths for all possible pairs of network nodes.

Indeed, it implies that the most important citation appears in 13.54% of all the paths found. The trajectory links patents from the end of the 1970s to 2013.

The main path is composed by 17 patents, with SPLC values varying from 0.0021–0.0402. Nodes composing the main path present different levels of in-degree from zero (source node) to 66 and k-core varying from 5–13. After the inspection of the patents, submitted to a panel of experts, it became clear that the main path was related to other lignocellulosic sources, like waste, to produce ethanol.

There are five patents that are crucial to the evolution of the field in terms of technological trajectories, that can be classified as "industrial processing."[14] What is striking is that trajectories with bigger SPLC records—linking patents in origin to those at the end of the network (more recently filled) pass through these five patents related to "how to prepare" materials to be processed and not related to the use of sophisticated techniques of molecular biology or enzymatic hydrolysis.

Fig. 8.3 illustrates the idea of technological trajectory into a giant component. The blue line represents the connection between patents filed in 1980 and patents recently filed: from general processes to produce ethanol to processes related to the use of waste.

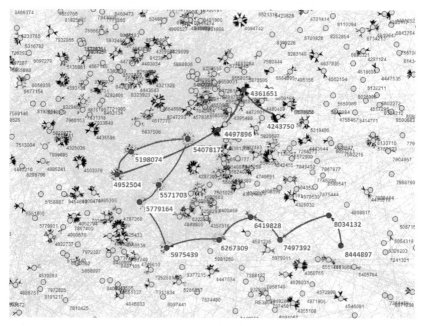

FIGURE 8.3 Technology trajectory of bioethanol (1976–2012). *From USPTO, using Odissey and PAJEK.*

---

14. International Patent Classification (IPC) A23K—feeding stuffs especially adapted for animals; methods especially adapted for production thereof.

Inspecting a bunch of trajectories with different starters it is possible to conclude that at least in the field of USPTO there is a kind of break between what had been done during the 1980s and the emergence of new alternatives to produce bioethanol (second-generation, algae, synthetic biology), result that put at risk the applicability of the concept of cumulativeness, a main component of the TT. For example, main assignee identified at the end of trajectories (filed after the year 2000) like Mascoma, Inbicon, Iogen, Dupont, Monsanto, Solazyme, BP, Xyleco, University of São Paulo, Zea Chem, DSM, were not found in older patent records, but other organizations that were found demand a worth mention: Tate & Lyle, Genencor, Solvay, National Distillers, Alfa-Laval, General Electric, and many universities (Darthmouth College, Missouri, Ohio, Florida, MIT, Midwest Research Institute, Kansas).

A further exercise was an attempt to remove the influence of the five aforementioned patents in the configuration of TT. The result is a clear identification of a TT related to a very well defined and mature petrochemical route to produce ethanol, where Celanese, BP, Exxon and Marathon GTF Technologies are the main assignees.

FINAL REMARKS

Bioenergy and the new ways to produce bioethanol comprise a new paradigm still in progress. From the point of view of this chapter, it is necessary to investigate the sector in a broad perspective, avoiding making conclusions based on case studies, particularly those related to firms that apparently succeeded to innovate.

Two groups of complementary methodologies have been used to deal with big data analysis, since the science and technology and innovation (ST&I) indicators must consider thousands or millions of data units, as articles, patents and intellectual property events.

The first methodological approach is developed to capture broad scientific and technological competencies and technological fields relations, to map and compare the countries competencies to publish, to publish in collaboration and even to identify their propensity to patenting activities, among other means to assign property rights. This section showed how useful the use of scientific papers to identify players, connections and research in bioenergy can be, mostly when the technological paradigm is in progress.

It is possible to conclude: (1) there is a well-defined group of countries leading the scientific and technological research with an increasing number of publications and with a central role in the networks of scientific collaborations, with higher prestige: EUA and China. (2) A second group is comprised of countries that typically take the role of "collaborators," like Germany, Sweden and Finland. These countries have been establishing connections with the aforementioned leading countries. (3) Brazil has an important role in bioenergy and bioethanol, but the connections are strong with only two central countries, but weaker than countries that assume the role of collaborators.

Inspecting the role of organizations, again it is clear that the central role of some organizations of the leading countries, such as for the USDA, University of California, Chinese Academy of Sciences. In the top 5 organizations in the ranking of scientific production in bioenergy research field, University of São Paulo is the most important representative of the Brazilian scientific organizations. This organization links a local subnetwork with the leading organizations in the world. However, there is no clear evidence of a relevant subnetwork on sugarcane (or saccharum) subarea, where Brazil is an international leader (Bueno et al., 2015).

The second methodological approach is to build networks of patents connected by citations. The social network analysis (SNA) allows the interpretation of the whole network and the evaluation of the importance of selected vertices, applying measurements of formal relation algorithms that come from the graph theory (Freeman, 2004).

Differently from the results achieved from scientific networks, a TT of bioethanol is not clear. The exercise of building a network of patents resulted in a clear divide between efforts conducted before the 1990s and the emergence of the bioenergy and bioethanol sector after the second half of the first decade of the 21st century.

This result is confirmed by the study of Murakami et al. (2015) that shows, in a specific and emergent subarea of the research to produce second-generation bioethanol nonconvergent trajectories, with new combinations of methodologies, enzymes, microorganisms and raw materials in the aim of solving guide posts generated by recent attempts to implement technologies.

This apparently undefined situation of the paradigm is in contrast with the fact that there is a visible group of players that are enrolled in the growing activity of patenting in the aim of protecting specific assets that will be crucial to the diffusion of the technologies in the near future.

In spite of the fact that Brazil has an important presence in the scientific scenery, with at least one organization placed in the center of the network, the results have brought some doubts about the capacity of the local organizations to generate technologies diffused worldwide. The fragile networking of countries dealing with feedstock obtained from sugarcane puts a lot of weight on the shoulders of the Brazilian organizations surrounding the University of São Paulo. A prospective issue arises in the horizon: even being a relevant collaborator in the scientific scenery would Brazil be an important player in the future of bioethanol production?

ANNEX I

Research parameter for scientific publications in selected topics: keyword, title and abstract

Research Parameter
TS = (*ethan* OR *energ*) AND TS = (*sugar* OR *cane* OR bagas* OR straw* OR cogener*) AND TS = (*conversion* OR *lign* OR *cellul*) AND TS = (*hydrolys* OR *ferment* OR *enzym* OR fung* OR *bac* OR *pressur* OR steam* OR chem* OR sacch* OR microb* OR clostrid* OR thermocell* OR *spor* OR *cocc* OR erwinia* OR strept* OR sclerot* OR phaneroch* OR trichod* OR asperg* OR schizoph* OR *penicill* OR SCP OR "Single Cell" OR *xyl*)Databases = SCI-EXPANDED, SSCI, A&HCI, CPCI-S, CPCI-SSH Timespan = All YearsLemmatization = On

Source: de Souza, L.G.A., Moraes, M.A.F.D., Dal Poz, M.E.S., da Silveira, J.M.F.J. 2015. Collaborative networks as a measure of the innovation systems in second-generation ethanol. Scientometrics (Online), 103, 355–372. doi:10.1007/s11192-015-1553-2.
6053 papers until October 24, 2012.

REFERENCES

Bell, M., Pavitt, K., 1993. Technological accumulation and industrial growth: contrasts between developed and developing countries. Ind. Corporate Change 2 (2), 157–211.

Brown, R.C., Brown, T.R., 2012. Why Are We Producing Biofuels? Brownia LLC.

Bueno, C.S., Silveira, J.M.F.J., Antonio de Souza, L.G., Masago, F.K., Buainain, A.M., 2015. An evaluation of how Brazilian scientific research contributes to the patenting of sugarcane ethanol. In: 19th ICABR Conference, 2015, Ravello. Anais.

Dal Poz, M.E.S., Silveira, J.M.F.J., 2015. Trajetórias Tecnológicas do Etanol de Segunda Geração. In: Salles Filho, S. L.M. O Futuro do Bioetanol: o Brasil na Liderança, 1ª Ed, Editora Elsevier, Campus, Cap 6, (p111:125), p. 185.

Dal Poz, M.E.S., Silveira, J.M.J., Masago, F.K., 2013. Technology frontier on bioenergy: analysis of two networks of innovation. In: Globelics Conference, 2013, Ankara. Anais da Globelics Conference.

Ferrari, V.E., 2015. Seleção e apropriação de biotecnologias agrícolas: uma análise sobre as trajetórias tecnológicas associadas aos organismos geneticamente modificados. Tese (doutorado) – Universidade Estadual de Campinas, Instituto de Economia. Campinas, SP: [s.n.].

Fontana, R., Nuvolari, A., Verspagen, B., 2009. Mapping technological trajectories as patent citation networks. An application to data communication standards. Econ. Innov. New Technol. 18 (4), 311–336.

Freeman, L.C., 2004. The Development of Social Network Analysis: A Study in the Sociology of Science. BookSurge, North Charleston.

Frenken, K., 2006. A fitness landscape approach to technological complexity, modularity, and vertical disintegration. Struct. Change Econ. Dyn. 17 (3), 288–305.

Hall, B.H., Jaffe, A.B., Trajtemberg, M., 2001. The NBER Citations Data File: Lessons Insights and Methodological Tools. National Bureau of Economic Research Working Paper 8498. <http://www.nber.org/papers/w8498>.

Hummon, N. P. e, Doreian, P., 1989. Connectivity in a citation network: the development of DNA theory. Social Networks 11 (1), 39–63.

Jackson, M., 2010. Social and Economic Networks. Princeton University Press, Princeton.

Jaffe, A. e, Trajtemberg, M., 2002. Patent, Citations and Innovations. MIT Press, Cambridge.

Masago, 2012. Odysseýs sistema para análise de documentos de patentes. Dissertação (mestrado) – Universidade Estadual de Campinas, Instituto de Computação. Campinas, SP: [s.n.], 2013.

Murakami, T.G.L., Silveira, J.M.F.J., Antonio de Souza, L.G., Bueno, C.S., 2015. Applying entropy indexes to identify technological trajectories: second-generation bioethanol production. In: 43 Encontro Nacional de Economia (ANPEC), 2015, Florianópolis. Anais.

Nooy, W., Mvar, A. e, Batagelj, V., 2005. Exploratory Social Network Analysis with Pajek. Cambridge University Press.

Shibata, N., Kajikawa, Y., Takeda, Y., Matsushima, K., 2008. Detecting emerging research fronts based on topological measures in citation networks of scientific publications. Technovation 28 (11), 758–775. http://dx.doi.org/10.1016/j.technovation.2008.03.009.

Souza, L.G.A., 2013. Redes de inovação em etanol de segunda geração. Tese (doutorado) - Universidade de São Paulo, Escola Superior de Agricultura "Luiz de Queiroz". Piracicaba, São Paulo.

Souza, L.G.A., Moraes, M.A.F.D., Dal Poz, M.E.S., da Silveira, J.M.F.J., 2015. Collaborative networks as a measure of the innovation systems in second-generation ethanol. Scientometrics (Online) 103, 355–372. http://dx.doi.org/10.1007/s11192-015-1553-2.

Souza, R.F., Leite, S.C.F., Ballini, R., Silveira, J.M.F.J., 2014. Emerging areas of research on bioenergy. In: 15 International Joseph A. Schumpeter Society, 2014, Jena. 15 International Joseph A. Schumpeter Society.

Verspagen, B., 2007. Mapping technological trajectories as patent citation networks: a study on the history of fuel cell research. Adv. Complex Syst. 10 (1), 93–115.

Vorhees, E., 1994. Query expansion using lexical-semantic relations. In: Proceedings of ACM SI-GIR International Conference on Research and Develoment in Informations Retrieval, pp. 61–69.

Wagner, C.S., Leydesdorff, L., 2005. Network structure, self-organization, and the growth of international collaboration in science. Res. Policy 34 (10), 1608–1618. http://dx.doi.org/10.1016/j.respol.2005.08.002.

Chapter 9

China's Fuel Ethanol Market

H. Lu

3E Information Development & Consultants, 3-eee.net, Wayland, MA, United States

INTRODUCTION

The dazzling economic growth over the past three and half decades has been driving up oil demand in China greatly. Annual oil consumption reached 566.2 million tons (or 12.52 million barrels per day) in 2015, accounting for 13% of the world total. Among major refined products, transportation fuels have been among the strongest in the crude slate (gasoline and LPG for cars and motorbikes, kerosene as jet fuel, diesel for trucks, locomotives, buses and boats, and bunker fuel oil for ships) along with surging needs for materials and cargo shipping and people's traveling. Fig. 9.1 displays shares of individual major oil products in total oil consumption in China in 2014.

The sustained economic advancement has helped lift people's living standards. Per capita GDP in China went near US$ 8,000 in 2015, when average urban per capita disposable income climbed up above US$ 6500 in many large cities along the coast. This has led the country to enter the automobile era. By the end of 2015, the total number of civilian automobiles in China approached 172.3 million, of which nearly 144 million were privately owned motor vehicles. In Beijing alone, the number of cars privately owned per 100 households exceeded 50 by 2014. Driving has increasingly integrated with people's daily lives. As a result, gasoline consumption has soared dramatically. Gasoline demand growth was recorded at 6.8% on average compared to 6.2% growth of total oil consumption over the last 20 years. In particular, growth of gasoline demand has accelerated over the last few years, as car driving is becoming a routine daily activity of an increasing number of residents, particularly in the relatively developed regions. During 2014, when the economic growth slipped down to 7% and increases of total oil demand slowed down to just 2.3% (in weight), gasoline demand growth jumped up 11% (see Fig. 9.2).

With accelerating investment and efforts made in resource exploration, China has greatly raised its petroleum resource appraisal, proving its resources

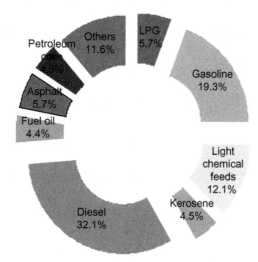

FIGURE 9.1 Oil consumption structure in China. *From 3E database.*

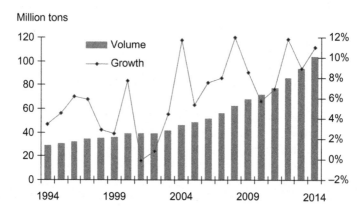

FIGURE 9.2 Gasoline consumption in China. *From 3E database.*

are not as poor as previously believed. The most recent round of national oil and gas resource appraisal conducted by the Ministry of Land and Resources put total conventional oil resources at 108.5 billion tons and recoverable crude oil reserves at 26.8 billion tons by 2013, 42% and 26% more than the previous appraisal done in 2007, respectively. The appraisal on gas resources was even encouraging. Total natural gas resources and recoverable reserves in the country climbed up to 68 trillion cubic meters and 40 trillion cubic meters, respectively, almost double the figures announced with the 2007 appraisal.

China is also endowed with unconventional oil and gas resources. In fact, several authoritative international organizations have estimated total shale gas resources could be more than 30 trillion cubic meters, while the Ministry of Land and Resources have made a recent assessment of 25 trillion cubic meters, the largest in the world with either number. In addition, shale oil resources are also quite abundant in China, with total resources of more than 47 billion tons and recoverable reserves of 16 billion tons. The central government has offered a subsidy of 0.4 Yuan per cubic meter to encourage shale gas development, and opened two rounds of shale gas exploration license tendering. However, like the conventional gas industry, the shale gas sector remains dominated by CNPC and Sinopec with a limited number of other investors. By 2015, a total over 25 billion Yuan investment was made in the sector for 900 wells and accumulated shale gas output of 4.5 billion cubic meters. An ambitious state long-term shale gas plan has been made to establish annual shale gas producing capacity of 6.5 billion cubic meters by 2015 and 30 billion cubic meters by 2020. However, the current relatively low oil prices in addition to complicated geophysical conditions, unfavorable geographical locations and insufficient water supply are posing enormous challenges to unconventional resource development in China.

In spite of enormous efforts made by domestic producers, indigenous oil production capacity expansions have been much dwarfed by corresponding increases in domestic oil demand. China has become a net oil importer since 1993, and is now dependent on foreign oil for around 60% of its total oil supply.

With large volumes of oil imports, the Chinese oil sector has become quite closely integrated with the global oil market. Domestic gasoline and diesel prices are adjusted every 10 working days based on a basket of international crude oil prices, though the adjustment remains in the hands of the central government. Jet fuel prices are also linked to the international market subject to monthly changes. The other refined product markets are generally opened with prices mostly market-oriented. In terms of oil product taxation, in addition to 17% value-added tax, consumption tax of 1.40 Yuan per liter is levied to gasoline, naphtha, solvent and lube oil; and 1.10 Yuan per liter to kerosene, diesel and fuel oil (jet fuel is temporarily exempted from the consumption tax).

DEVELOPMENT OF FUEL ETHANOL IN CHINA

China began developing fuel ethanol in the early 21st century. As the country's oil import dependence was quickly ascending, and urban air pollution was getting increasingly severe, development of alternative and cleaner fuels began to rise to the top of the fuels agenda in China. Then, growing overstored grain reserves after years of agricultural harvests, becoming stale and possibly decreasing in value, happened to create a favorable circumstance and acted as a direct driver to the development of ethanol fuel in China. Fuel ethanol produced from grain seemed just right in responding to all the above concerns.

Development of fuel ethanol was launched and guided by central government, representing a country-level strategic action. The State Council and major governmental authorities issued a series of administrative orders and notices on fostering the industrial chain of fuel ethanol production and applications from formulating essential industrial standards, amending related regulations, building fuel ethanol manufacturing plants, setting up blending and distributing centers and establishing trial application zones.

Four fuel ethanol manufacturing plants were established, namely Huarun Ethanol in Heilongjiang Province, Jilin Fuel Ethanol in Jilin Province, both located in northeast China; Tianguan Fuel Ethanol in Henan Province in central China, and Fengyuan Bio Chemical in Anhui Province in east China, with total phase-one capacity of 1.02 million tons per year. The required ethanol–gasoline blending centers and distribution channels were built by CNPC and Sinopec in designated market areas. The two state-run oil companies were the dominant oil players in the northern and southern halves of the domestic oil market respectively, and had established complete and extensive logistic systems of oil product storing, distributing and marketing in their market territories. This helped facilitate quick promotion of ethanol–gasoline applications in the market.

In February 2004, after trial experimentation of E10 ethanol gasoline (10% fuel ethanol blending in gasoline by volume) for nearly 2 years in a few cities in Henan Province and Heilongjiang Province, the central government officially ordered the expansion of ethanol–gasoline applications in the country. Full ethanol–gasoline utilization was applied to five provinces of Heilongjiang, Jilin, Liaoning, Henan and Anhui; and partial utilization to four other provinces of Hebei, Shandong, Jiangsu and Hubei. Later, along with the construction of additional fuel ethanol plants, Guangxi Autonomous Region in southwest China joined the full application zones in 2008, and Inner Mongolia Autonomous Region in north China became another partial market zone in 2014 (see Fig. 9.3).

Total fuel ethanol utilization has reached 2 million tons annually, translating into 20 million tons of ethanol–gasoline blend, accounting for around 20% of the gasoline market in China (Fig. 9.4).

The first four fuel ethanol plants all use food grain as the feedstock with corn in Huarun Ethanol, Jilin Fuel Ethanol and Tianguan Fuel Ethanol; and wheat in Fengyuan Bio Chemical. The common conversion of starch to fermentable sugars and fermentation technology is applied. China Oil and Food Corporation Group (COFCO) constructed a cassava-based 200,000 tons per year fuel ethanol plant in Beihai of Guangxi Autonomous Region, and Zhongxing Energy in Inner Mongolia built an annual 100,000-ton fuel ethanol plant taking sweet sorghum as the feedstock. The most recent, Longli Bio Fuel, has turned the country's first cellulosic ethanol plant into commercial operation with an annual fuel ethanol production capacity of 50,000 tons in Shandong Province (see Table 9.1).

China's fuel ethanol programs have been led and tightly controlled by the central government. The National Development and Reform Commission

FIGURE 9.3   China's fuel ethanol program map. *From 3E database.*

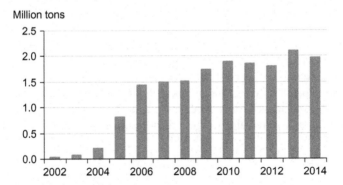

FIGURE 9.4   Fuel ethanol utilization in China. *From 3E database.*

(NDRC, previously State Planning Commission) supported by a few other authoritative organizations has taken the lead in issuing governmental orders, mandates and policy measures to guide the implementations of the programs. A series of preferential policy measures and financial aid are provided to the program players as designated by NDRC. While fuel ethanol

TABLE 9.1 Fuel Ethanol Manufacturing Plants in China

Plant	Location	Feedstock·	Capacity (1000 tons/year)
Huarun Ethanol	Heilongjiang	Corn	250
Jilin Fuel Ethanol	Jilin	Corn	600
Tianguan Fuel Ethanol	Henan	Corn	500
Fengyuan Bio Chemical	Anhui	Wheat	440
COFCO Beihai Ethanol	Guangxi	Cassava	200
Zhongxing Energy	Inner Mongolia	Sweet sorghum	100
Longli Bio Fuel	Shandong	Cornstalk	50
Total			2140 (±2 billion liters)

Data source: 3E database.

manufacturing plants are invested in and operated by quite a few state-run companies, ethanol–gasoline blend distribution and sales are exclusively conducted by CNPC and Sinopec.

From the very beginning, a wide range of advantageous measures were designed to support fuel ethanol development and utilization, including special grants and low-interest loans for investment in fuel ethanol plants and ethanol–gasoline blend infrastructure, subsidized grain feedstock to fuel ethanol plants with prices set lower than commercial market prices, exempting ethanol gasoline from consumption tax generally levied on auto fuels, and refund of value-added tax to fuel ethanol players. In addition, direct subsidies have been provided by the Ministry of Finance for fuel ethanol plants, as fuel ethanol manufacturing costs in China have remained higher than gasoline for the majority of recent years.

Fuel ethanol blending, distribution and other logistics involve extra costs to CNPC and Sinopec. To compensate, the central government sets prices of fuel ethanol delivered from ethanol plants to CNPC and Sinopec lower than corresponding gasoline ex-plant prices. Fuel ethanol deliveries are priced at 91.11% of #90 gasoline ex-plant prices. Ethanol gasoline is sold at gas stations at the same price as equal-grade gasoline. Subsidies for fuel ethanol production were at the beginning calculated with costs plus minimum profits of fuel ethanol plant operations, and since 2006 have become fixed amounts adjusted each year following the movements of domestic gasoline prices, which are linked to global oil prices. Since 2012, central government has introduced a phasing-out

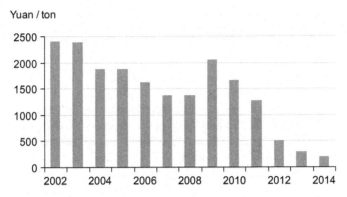

Yuan / ton

FIGURE 9.5   Subsidies to fuel ethanol in China. *From 3E database.*

mechanism, and subsidies for grain-based fuel ethanol manufacturing plants were gradually reduced to 0 by 2016 (Fig. 9.5).

FUEL ETHANOL POLICIES FOR THE FUTURE

Despite the initial success of fuel ethanol utilization as a cleaner renewable fuel to reduce traditional fossil fuel usage and cutting auto emissions, some of its deficiencies were gradually exposed in the course of its practical applications after an initial success in implementing the central government mandatory fuel ethanol programs. Although many of those deficiencies are relatively minor problems in technologies, logistics, administration and cross-sector cooperation, such as lower heat value of ethanol, erosions caused by ethanol though relatively minor, water absorption of ethanol causing water–fuel separation, impacts on engine power and acceleration, etc. Many of these problems remained unresolved as no clear responsibility of technical support was assigned, gradually discouraging and even driving industrial players and consumers away.

Moreover, sustainability of feedstock supply and economics of manufacturing fuel ethanol have posed heavy challenges to the future development of fuel ethanol programs. As the grain market in China turned from oversupplied to insufficient within just a few years during the early 2000s after the introduction of fuel ethanol programs, the central government was forced to issue an urgent order in late 2006 to ban any new grain-based fuel ethanol investment except for the existing first four fuel ethanol plants, turning to encourage uses of non-grain agricultural products such as cassava, sweet sorghum and sweet potato as feedstock; and development of a new generation of cellulosic technologies based on agricultural wastes such as crop stalks and corncobs. However, as collection and storage of those alternative agricultural feedstocks are not yet well established, much of the cassava feedstock for COFCO Beihai fuel plant in Guangxi Autonomous Region has in fact to rely on high-cost imports from

Southeast Asian countries, and sweet sorghum-based Zhongxing Energy fuel ethanol plant in Inner Mongolia and other smaller trial projects are facing insufficient feedstock supply.

In addition, the phasing-out subsidies for fuel ethanol manufacturing and sharply plummeting oil prices have greatly squeezed fuel ethanol plant operating margins and pushed their operations deep into the red. The combination of falling enthusiasm of ethanol–gasoline blend consumers and discouraged fuel ethanol producers (mainly because of oil prices) has resulted in declines in total fuel ethanol production and ethanol–gasoline blend utilization, which would likely continue into 2015. For example, the full ethanol–gasoline blend market in Guangxi Autonomous Region has gradually been eroded by local gasoline suppliers with discounted regular gasoline. In Shandong Province, a growing number of auto drivers went back to burn normal gasoline instead, forcing gas filling stations in full ethanol–gasoline blend market zones to stop ethanol–gasoline blend services and return to regular gasoline business. Similar situations are also seen in the other full ethanol–gasoline blend market zones. In mid 2014, Anhui Province—one of the first five full ethanol–gasoline blend utilization provinces—took action to challenge the central government's designated full ethanol–gasoline blend market zones, and reintroduced regular gasoline in a few gas stations in Hefei, the provincial capital. Some local governments and organizations have also submitted proposals to central government, requesting to stop fuel ethanol programs.

On the other hand, central government has in general remained positive in the policy orientation to continue supporting fuel ethanol development in spite of the decision of phasing out subsidies for fuel ethanol production. Development and utilization of fuel ethanol have been included in a series of governmental plans and state energy strategies. The long-term renewable energy and biofuel development plans have set aggressive goals of annual fuel ethanol utilization reaching 4 million tons by 2015 and 10 million tons by 2020, thought the goals now seem unreachable. In 2012, NDRC gave approvals to another five new fuel ethanol manufacturing projects utilizing cassava, sweet sorghum and sweet potato as feedstocks.

After three decades of rapid economic expansions, China has entered a new stage of development with growth rates evidently moderating in both the economy and energy consumption. However, compared to other energy sources such as coal, electricity and even natural gas, oil demand has stayed rather strong. In spite of slowing down consumption of diesel, naphtha and fuel oil with the impacts of industrial production becoming sluggish, demand for majority of transportation fuels including gasoline, jet fuel, auto diesel and bunker fuel has remained relatively healthy. Soaring gasoline consumption is even more outstanding. The total number of automobiles in China is expected to more than double within the next 10 years. This will sustain the strength of the gasoline market in the coming years. Meanwhile, as increases in projected indigenous oil production trail behind corresponding domestic oil demand growth, China's oil import dependence is expected to continue rising. Forecasts have shown that Chinese refiners will receive increasing volumes of sour and heavy crude oil from the Middle East, South American and Central Asian producers and Russia as well. The unfavorable crude oil slates combined with tightening environmental regulations and auto fuel standards will add pressure and challenges on domestic refineries. To fight intensive and extensive air pollution throughout the country, the central government has repeatedly urged advancement of the implementation of auto fuel quality upgrading plans. The state fifth-stage gasoline and diesel standards (equivalent to the Euro 5 auto fuel standards) will become effective in all 11 coastal provinces from early 2016, and then country-wide at the beginning of 2017 (ethanol gasoline standards would likewise be upgraded).

Promoting renewable energy and clean auto fuels has become a key section of the country's long-term strategy and plans for sustainable development. In parallel with fuel ethanol development, a few other cleaner fuel programs, mainly natural gas, biodiesel, fuel methanol and electric cars have also been under development in China with support from various levels of governmental authorities.

A national campaign for natural gas development and utilization is also extended to gas-fueled automobiles, being mainly stimulated by comparatively low domestic natural gas prices. As the compressed natural gas and liquefied natural gas (CNG/LNG) market in China is relatively liberalized, the development of gas-fueled automobiles have quickly drawn in a wide range of investors and players, helping foster the full industrial chain from CNG distribution, inland small LNG manufacturing plants, LNG shipping, CNG/LNG filling stations and car retrofitting and gas-fueled automobile manufacturing. Many local governments have provided preferential policy measures and financial support to gas-fueled automobile development. In spite of lifting natural gas prices in recent years, the central government has set CNG/LNG prices to link to 85% of corresponding gasoline prices for automobiles, and 60% for public service buses and taxies, providing good operating margins of gas-fueled automobile business in general. Estimates have shown that the total number of gas-fueled automobiles (excluding liquefied petroleum gas (LPG) cars and motorbikes) exceeded 4.2 million, and near 7000 CNG/LNG filling stations had been established spreading in almost all the provinces of the country by 2015.

The central government's initiative to promote biofuels has also encouraged biodiesel development. Many private investors have been involved, as the market is generally opened for social investment. In an estimate, over 2 million tons per year of biodiesel production capacity has been constructed in China. Immediate feedstock sources for biodiesel are waste vegetable cooking oil and animal fat. Some investors have also tried to grow oil-rich plants, but initial projects of this kind seem unsuccessful. Insufficient feedstock supply, poor fuel quality and resistance of the mainstream auto fuel retailing chain to accept biodiesel are among the main obstacles. So far, less than a half million tons of biodiesel each year are produced in China, the majority of which are however used as an industrial fuel. CNOOC's annual 600,000-ton biodiesel plant in Hainan Province currently running with imported palm oil is the only biodiesel project in the country officially delivering biodiesel into local auto fuel filling stations.

The majority of methanol in China is produced from coal. Over the years, a few coal-rich provinces have made efforts in promoting fuel methanol in auto fuels. Shanxi, Shaanxi and Guizhou provinces have formulated local fuel methanol standards. Although methanol supply is quite abundant and the costs are low, reluctant acceptance of consumers, opposition stance of major state oil companies and unclear policy tones of central government have placed restraints over fuel methanol development and utilization. Shanxi Province has conducted years of trial programs of M15 methanol gasoline (blending 15% methanol in gasoline) in a few cities, but repeatedly postponed an ambitious plan of extending methanol gasoline into the entire province. After initial experiment of M15 and M25 methanol gasoline in its main cities, Shaanxi Province has also canceled its plan to promote a province-wide enclosed methanol gasoline application. Since 2012, the Ministry of Industrial and Information Technology launched high-percentage methanol fuel (M85 and M100) experiment programs in Shanxi Province, Shanghai and Shaanxi Province, which were later expanded into Gansu and Guizhou provinces. The market is now waiting for the results of the experiment programs and the decision of the Ministry on the future development and utilization of fuel methanol.

Being eager to catch the global trend and play a leading role in the new generation of advanced technologies, China has made aggressive plans to develop electric cars with goals of accumulated electric cars (electric and plug-in hybrid) arriving at 500,000 by 2015, and annual electric car production reaching 2 million by 2020. However, the early-stage development was extremely disappointing. By 2013, total electric cars in China were much below 100,000, far lagging behind the state plans. To create more incentives to the sector, the central government has launched a new campaign offering large amounts of financial support including direct subsidies and tax and fee exemptions. Being greatly stimulated, annual electric car production in China went over 80,000 in 2014, and 340,000 in 2015. However, the trends of electric car development in China that have been mostly relying on the governments' heavy subsidies will be uncertain.

TRENDS AND POLICY IMPLICATIONS

The prospect of future development and utilization of fuel ethanol in China looks rather mixed under the ongoing economic, energy and environmental context and the overall governmental policy framework. To some extent, the slowing down of economic growth and energy usage is relieving the pressures of energy supply security and CO_2 emissions. Since China made a commitment in late 2014 to cap CO_2 emissions by 2035, there has been no progress seen in issuing further policy details. Most recently, the top policy priority of the Chinese government has shifted to boost investment in industries and infrastructure to regain growth momentum to the economy and to maintain social stability, though the policy tone of environmental protection remains harsh. On the other hand, regulations and efforts over reducing urban air pollution are strengthened, and promoting efficient and cleaner energy including auto fuels is still one of the governments' chief agendas.

Annual gasoline consumption in China is exceeding 100 million tons (2.3 million barrels per day), offering large market potentials for cleaner fuel utilization. However, several alternative cleaner auto fuels have already been introduced into the market, and among them, fuel ethanol's advantages in reducing auto emissions might not be exceptional. In theory, ethanol gasoline is cleaner than normal gasoline, but in practice, the results of applying ethanol gasoline (E10) are not so clearly measured and recognized. Nevertheless, in spite of an increasing number of provinces' adoption of ethanol–gasoline blends, major severely polluted metropolitan cities including Beijing, Tianjin, Shanghai and Guangzhou have not yet adopted fuel ethanol. They are instead tending to introduce gas-fueled and electric cars. Central government's policy orientation of new-energy automobiles is seen exclusively shifting to supporting electric cars (electric and plug-in hybrid). The future fuel ethanol development and utilization are gradually left for commercial operations and market competition except for the government controls of grain-based fuel ethanol investment.

In terms of feedstocks, constraints of supply may expect gradual improvement in both grain and nongrain feedstocks. It is reported by authoritative agricultural and food organizations that as many as 12 million tons of grain (annual grain production in China is around 600 million tons) produced each year are unqualified as food with high heavy-metal contents exceeding the state standards due to severe soil pollution. Industrial experts have already made proposals to use the so-called "problematic grain" as feedstock for fuel ethanol manufacturing. For the category of nongrain agricultural products, there are huge areas of noncultivated land (far more than needed fuel ethanol manufactured from nongrain agricultural products) including saline-alkali land, waterlogged low land, hillside fields and other wasteland throughout the country that is not suitable to grow grain but good to plant cassava, sweet sorghum and sweet potato. The future of nongrain agricultural product-based fuel ethanol development will lie in perfect integration of agricultural and industrial operations and cost-effective feedstock gathering, storing, preprocessing and shipping, which are also crucial to cellulosic fuel ethanol. These, together with technology development, will be determinant forces for future fuel ethanol development. Externally, movements of global oil prices will be decisive as to how fast and how early fuel ethanol could replace normal gasoline and penetrate the market. Cellulosic technology development was heated when oil prices soared three-digits high, but is seen to be cooling down with plunging oil prices. The majority of the existing fuel ethanol plants have made active investments in cellulosic research. Eight experimental cellulosic ethanol plants across the country have been built or are under construction, with a total fuel ethanol production capacity of around 60,000 tons per year.

FINAL REMARKS

In summary, fuel ethanol development in China is experiencing a transition from guided projects led by central government to a more commercialized course after governmental subsidies are gradually phased out. Technology development and costs will be the key factors to help fuel ethanol grow competition edges over normal gasoline and other alternative fuels.

In terms of potential fuel ethanol imports, China imports raw ethanol materials of nearly 40 million tons each year, mainly for alcohol and chemicals. The Chinese government has recently lowered the import tariff of ethanol from 30% to 5%, favoring additional ethanol imports. While opportunities of fuel ethanol imports do exist, the prospect should not be over optimistic. In the short to medium term, the troubles domestic fuel ethanol manufacturers have confronted could favor potential import, but the current ethanol–gasoline blend market conditions are not open for additional fuel ethanol supply. In the longer term, domestic production and supply could however likely become more relevant with all the expected improvements discussed above in feedstock supply, integrated agricultural–industrial operations and technologies being achieved. Under the above premises, fuel ethanol could yet prevail in the alternative auto fuel market in China in the future.

RELEVANT WEBSITES

<www.ndrc.gov.cn> National Development and Reform Commission of China.
<www.miit.gov.cn> Ministry of Industry and Information Technology of China.
<www.mlr.gov.cn> Ministry of Land and Resources of China.
<www.cpcia.org.cn> China Petroleum and Chemical Industrial Federation.
<www.CNPC.com.cn> China National Petroleum Corporation.
<www.Sinopec.com> China Petrochemical Corporation.

Chapter 10

Fuel Ethanol in Africa: A Panoramic View With an Accent on Bénin[1] and Kenya[2]

S.C. Trindade
SE²T International. Ltd, Scarsdale, NY, United States

INTRODUCTION

The global long-term future of biofuels lies in Africa for its geographical location, resource endowment and increasing energy services needs driven by development and population growth (UN, 1991). However, in the short term, there has been limited market penetration of biofuels, ethanol and biodiesel, in Africa's energy systems.

Updated information about fuel ethanol developments in Africa is not easy to come by, as web searches reveal.

There have been starts and stops in many countries' ethanol fuel programs, such as in Bénin (Trindade et al., 2012), Ethiopia (Agra), Kenya (Business Daily Africa), Malawi (EthCo), Mozambique (Leal, 2016; Ecoenergy, 2008), Nigeria (Nigeria), Senegal (Senegal), South Africa (South Africa, 2015; Nigerian National Petroleum Corporation Researchomatic, 2013), Zambia (Zambia) and Zimbabwe (Zimbabwe, 2014; Scurlock et al., 1991). Feedstocks that have been considered are sugarcane, maize, sweet sorghum and cassava (Batidzirai, 2007).

---

1. In 2007, the author was retained by ETA Florence to support a study on the strategic feasibility of biofuels in Bénin.
2. In 1982, the author was retained by UN/TCD to review the institutional framework in Kenya: the raw materials for alcohol production; the current power alcohol projects; the pricing of alcohol, gasoline and alcohol–gasoline blends; and standards and norms applicable to fuel ethanol. In addition to visiting Kenya in August 1982, there was also a study visit to Malawi, which had recently started using fuel ethanol.

Global Bioethanol.

Besides fuel for transportation, immediate markets for ethanol in Africa include domestic and export destinations. Ethanol is consumed as industrial ethanol, pharmaceutical ethanol, potable ethanol and ethanol cooking fuel gels. The latter product is used especially in Mozambique, where CleanStar Mozambique made it originally from cassava grown by local farmers in Dondo, but may now be outsourcing ethanol. A comprehensive analysis of the propects of biofuels in Mozambique was carried out by Ecoenergy and sponsored by the World Bank and the Italian Embassy in Maputo and published in 2008 (Leal, 2016). In Malawi, sugarcane ethanol has been made and used as a 10% blend (E10) since 1982 to extend gasoline in transportation markets. The two producers in Malawi are Ethanol Company (EthCo), in Nkhotakota District, and Presscane Corporation, in southern Malawi.

In Zimbabwe, a mandatory of E5 has been in place since August 2013. A voluntary E85 blend is also available. The sole supplier is Green Fuel located at Chisumbanje Estates, Manicaland Province. An older plant at Triangle, owned by South African interests, makes ethanol for export but occasionally may shore up supply from Green Fuel.

In South Africa, a mandatory E2-E10 blend is planned to come into effect by October 2015. However, the recent sharp fall in oil prices and the regulatory uncertainty about the subsidies available are putting investors on hold (South Africa, 2015b).

However, the focus of this chapter is on fuel ethanol in Bénin and Kenya where the author has had first-hand experience from his past work in the field. The material presented here is adapted from a previous publication of the author and has a historic flavor for the development of fuel ethanol in Africa.

FUEL ETHANOL IN KENYA

Fuel ethanol, then called power alcohol, was not the priority issue in Kenya's energy agenda in the early 1980s. The energy problem in Kenya was first and foremost one of supplying adequate firewood to the increasing residential demand derived from the country's very high population growth rate.

Second, Kenya's main energy problem related to displacing fossil fuel consumption in electricity generation, as well as supplying (or finding alternatives to) the fast-increasing incremental demand for automotive diesel oil. Alcohol substitution for gasoline therefore represented mainly a short-term opportunity for petroleum substitution and self-reliance.

Sugarcane has been cultivated in Kenya's highlands and could, in principle, provide the molasses required for conversion into fuel ethanol. This is why the three existing projects in 1982, in Muhoroni (Agro-Chemical and Food Company—ACFC), Kisumu (Kenya Chemical and Food Corporation—KCFC) and the Riana project were all located in the highlands area. The availability and low opportunity cost of the raw materials, in the form of molasses or sugarcane juice, made inland locations with limited logistical infrastructure especially attractive (Trindade, 1982).

Table 10.1 summarizes the basic facts about the Kenyan ethanol projects in 1982; in the case of ACFC and KCFC, the expected markets for other coproducts contributed to overall economic feasibility, which is in general a key feature of sugarcane, due to the many coproduct options available.

TABLE 10.1 Summary of Kenyan Ethanol Projects, 1982

Item	ACFC	KCFC	RIANA
Location	EASI, Muhoroni	Kisumu	South Nianza
Product mix	Ethanol	Ethanol	Ethanol
	Baker's yeast	Yeast	
	Fodder's yeast	Citric acid	
	Concentrated stillage	Methane	
	Fusel oil	Gypsum	
		Ammonium sulfate	
Ethanol capacity, m³/year	18,000	20,000	45,570
Investment per daily capacity, $/liter/day	211	383	437
Equipment provider	Vogelbusch	Conger/PEC/ SORIGONA	FCB/ SODECIA
Capital structure	ADC, ICDC (28% each)	GoK (<51%)	TJ Cottington
	IIC (34%), VEW (10%)	Chemfood Invest and Advait (49%)	OTH Int'l SUDE

Sources: from Trindade, S.C., 1982. Review of Power Alcohol Issues in Kenya. Report on Integration of Power Alcohol Operations with the UN/TCD Project KEN/76/005. Energy Planning, Policy and Programming in Kenya and Baraka, M., 1989. Alternative Energy for Transport Industry – The Kenyan Experience with Ethanol. Internal Document, Sales and Technical Services, Caltex Oil (Kenya) Limited.
ACFC, Agro-Chemical and Food Company; ADC, Agricultural Development Corporation, a Kenyan parastatal; Chemfood Investment Corporation, SA and Advait International, SA, Kenyan corporations; Conger, a Brazilian distillery manufacturer; EASI, East African Sugar Industries Ltd, sugar factory adjacent to ACFC; FCB, French sugar processing equipment manufacturer; GoK, Government of Kenya; ICDC, Industrial and Commercial Development Corporation, a Kenyan parastatal; IIC, International Investment Corporation, a foreign financing company; KCFC, Kenya Chemical and Food Corporation; OTH, French consulting and construction firm; PEC, Process Engineering Company, Zurich; SODECIA, French distillery equipment manufacturer and engineering company; SORIGONA, Swedish equipment manufacturer; SUDE, Sugar and Development, French implementation company; TJ Cottington & Partners Ltd, Kenyan engineering and industrial manufacturing company; VEW, Vereignigte Edelstahlwerke AG, Vogelbusch's parent company; Vogelbusch, Austrian distillery manufacturer.

Sugarcane alone was not an adequate sole source of fuel ethanol in Kenya. Any consideration of a truly national fuel ethanol program would have required, depending on the scenario considered, between two and five additional distilleries by 1995, which did not necessarily need to be based entirely on sugarcane (Trindade, 1982). Cassava and sweet sorghum cultivated on marginal lands would have been likely alternatives to water-demanding sugarcane agriculture. Assessment and needs identification were, to say the least, inadequate to support a national fuel ethanol program in Kenya.

In 1982, the KCFC complex was already stalled due to massive cost overruns. The fundamental error was to site a molasses-based manufacturing system, where there was no readily available molasses! Furthermore, there was no sugarcane bagasse to fire the boilers to produce process steam and electricity for the plant and perhaps for the commercial electrical grid.

Many years after, Spectre International, a South African Company, entered the process and took control of the assets in place in Kisumu in 2003. Energem Resources, a Canadian company, formerly known as Diamondworks Ltd., with many activities in Africa, including diamond mining, acquired for $2 million a 55% controlling interest in the Kenyan Spectre, to expand business opportunities in the Kenyan market. The distillery was revamped at some cost. By mid-2004 the distillery was ready to go and by 2008 was producing some 75,000 L/day (Crilly, 2008) mainly for the pharmaceutical and potable ethanol markets.

It seemed that, decades after its inception and a 10-year experience (1983–93), Kenya was likely to bring fuel ethanol back into use again. Indications were that from September 2010, the Kenya Pipeline Company's depots in Eldoret, Kisumu and Nakuru would begin blending 10% ethanol with gasoline. But by 2010, to meet the national requirement of 10% ethanol blend, Kenya's national production would have needed to triple! Currently, it seems that the two largest producers of ethanol in Kenya, Spectre International at 27 million liters/year in Kisumu and ACFC in Muhoroni, with 22 million liters/year are supplying the domestic potable and pharmaceutical markets and exporting ethanol.

A process of privatization of the five sugar mills, state-owned or -controlled, was expected by 2012, which would have affected the fuel ethanol prospects. However, by May 2015, the process was likely to take a further 9–12 months! In addition, the government was considering diversification of the millers into ethanol distillation and power cogeneration (CNBCAFRICA, 2015). Mumias Sugar Company had plans for a 100,000 L/day, US$ 45 million molasses-based fuel ethanol distillery due to start up in 2012 (Odhiambo, 2009). In August 2009, Mumias began generating electricity from bagasse-fueled boilers.

"We expect the first products off the shelf by July 2011. We anticipate to produce a variety of products including neutral alcohol, anhydrous ethanol, food grades and ethanol to blend with premium fuel," Evans Kidero, Mumias managing director, said (Odhiambo, 2009). Half of the distillery project funds would be drawn from the miller's retained earnings while the rest would be sourced from financiers. The distillery was scheduled to run on the 100,000 tons of molasses Mumias produced every year, and had a targeted production of 25 million liters of ethanol per annum.

"We are particularly eyeing the business area where ethanol would be blended with premium fuel in the ratio of 85:15. We expect the Ministry of Energy to give a comprehensive position on this and we shall take up the business opportunities," Kidero said at the time (Odhiambo, 2009). This statement confirms that government policy still was not sufficiently firm in mid-2010! "The market for ethanol is growing and the strategy of the proposed venture is to utilize all the molasses from the sugar plant for production of ethanol to be sold locally and regionally in order to exploit an opportunity in the market," he said.

Other projects announced in 2009 included Chemelil, Nzoia and Sony sugar factories to jointly produce up to 140 million liters/year of ethanol (Anonymous, 2010). From the information available presently, these projects appear not to have come to fruition.

Kenya is no exception to the requirement for learning time to develop and accumulate experience in implementing a new energy source such as fuel ethanol. The difficulties encountered—institutional, technical and otherwise—in implementing a fuel ethanol program are an integral part of this learning process. An awareness of such a "learning curve" helps expedite progress towards the market penetration of fuel ethanol.

It seems that limited learning has taken place over the past 30 years or so, and there is still much to do to achieve a consistent national fuel ethanol program in Kenya. The reality is that such a program must be discussed and agreed among the relevant stakeholders and translated into legislation and regulations. And such process has not been achieved in Kenya thus far.

A crucial development, which impacted on the prospects of ethanol production in Kenya was the ending in 2011 of the preferential trade terms on sugar with other producers within the Common Market for Eastern and Southern Africa (COMESA). This would leave local millers open to all-out competition (Endelevu, 2008; Odhiambo, 2009). However, a COMESA Council decision in March 2015 granted Kenya, for 1–2 years, an extension of the sugar safeguard (Wosemo, 2015).

However, to enjoy this benefit, Kenya must comply with certain requirements. These included "privatising state owned mills, doing research into new early maturing and high sucrose content sugar cane varieties and adopting them, paying farmers on the basis of sucrose content instead of based on weight, maintaining the safeguard as a tariff rate quota with the quota increasing while the above quota tariff falls until it reaches 0% and maintaining and providing infrastructure including roads and bridges in the sugar growing areas." (Wosemo, 2015)

FUEL ETHANOL IN BÉNIN

The case of Bénin is a study that contrasts with the case of Kenya. The learning curve in Kenya has run for at least 30 years, whereas in Bénin the curve is just beginning. The approach towards a national biofuel program in Bénin has been slow, but systematic. It has paid attention to the views of the relevant stakeholders and taken a holistic view of the country's biofuel prospects. Bénin does not yet have a national biofuel program, although it has made advances, including on the legislative front, with the creation in 2008 of a National Commission for the Promotion of Biofuels and a set of rules for companies interested in producing biofuels in the country. The development of such a program in Bénin has been supported by the World Bank system. Under this umbrella, the government decided to run a competitive bidding process for selection of a consulting group to conduct basic studies to inform its decision-making process. ETA Florence of Italy was selected to carry out such studies, from 2007, with the participation of the author of this paper and others. ETA Florence has discussed its reports (ETA, 2009a,b) with the government and has participated in Bénin in a round of stakeholders' dialogues. Still, in early 2011 there were no liquid biofuels produced or used in Benin as the country travels through the early steps of the biofuel learning curve.

The Republic of Bénin has a surface area of 114,763 km^2 and a population of 7,833,744. As with many other developing countries, agriculture plays a vital role in the national economy, employing nearly 75% of active populations and contributing 38% to GDP.

Given the country's growing population and developing economy, energy demand is rising. Transport in particular accounts for 23% of the total demand, with an average growth rate of nearly 11% per year for gasoline and 7% per year for diesel (average 1999–2005).

Even though some recent oil finds have shown the potential of offshore oil reserves within the territorial sea of Bénin in the Guinea Gulf, the current demand for fossil fuel is completely satisfied by imports. In a scenario of rising oil prices, the country's expenditure for energy provision would pose a serious constraint to its economic development.

In this context, the introduction of mandatory biofuel blends in conventional fuels sold at service stations could help to reduce fossil fuel imports. The development of a national biofuel program in Bénin has occurred in the context of a major energy crisis for the country, characterized by continuous power shortages and constant increases in fossil fuel prices, in a country heavily dependent on oil imports.

Preliminary estimates showed that, under current growth rates for gasoline and diesel consumption, the introduction of E10 blends would have required production of nearly 90 million liters of ethanol in 2010 and more than 400 million liters in 2025. The introduction of biofuel blends would not only contribute to a reduction in fuel imports, but would also bring environmental benefits and provide a potential market for the development of a national agro-industrial sector, thus stimulating rural development and diversification of agricultural activities.

In terms of export potential, Bénin could take advantage of the Cotonou Agreement signed in 2000 amongst USA/EU and ACP countries (Africa,

Caribbean, Pacific), which would allow a total exemption of import tariffs for ethanol and biodiesel into the EU and the USA. Despite some limiting factors, the agricultural sector generally covers the country's food demand, the main food crops being cassava, maize, grain sorghum and groundnuts, whereas the two main cash crops for export are cotton and cashew nuts. The introduction of E10 blends would require the cultivation of 20,000–25,000 hectares of energy crops in 2010, increasing to 80,000–90,000 hectares in 2025.

Bénin's climate conditions vary among different regions, ranging from the sub equatorial climate of the south (1300–1500 mm of annual rainfall with two rainy seasons) to the semiarid conditions of the north (700 mm and one rainy season). As far as ethanol is concerned, three main feedstock options were identified: sugarcane (*Saccharum officinarum*), sweet sorghum (*Sorghum bicolor*) and cassava (*Manihot esculenta*). Sugarcane is certainly a suitable option for ethanol, well-known worldwide, which could yield up to 6000 liters per hectare if cultivated under the right conditions (preferably in the southern agro-ecological zones with a subequatorial climate and higher annual rainfall, as long as it does not displace food crops). Moreover, ethanol from sugarcane has a very positive energy balance as its lignocellulosic biomass (bagasse) can fuel boilers for heat and power generation and even export electricity to the grid. At present, 4000–5000 hectares of sugarcane are cultivated in Bénin in a single irrigated complex for the production of sugar.

Although sweet sorghum is not yet cultivated in Bénin, traditional grain sorghum (which has similar cultivation techniques and requirements) is well introduced into the farming system, being an important food and fodder crop, especially in the northern regions with semiarid conditions. Sweet sorghum could represent a very promising feedstock for ethanol production. Compared to sugarcane, sweet sorghum has a shorter growth cycle (4–5 months instead of 12–18 months for sugarcane); it is an annual crop, although under proper conditions it could be cultivated in two or even three cycles per year. In optimal conditions the potential ethanol yield of sweet sorghum is comparable to that of sugarcane (more than 5000 liters per hectare).

Cassava is widely cultivated throughout Bénin and is an important food crop for the country, especially in rural areas; it is estimated that at least 68% of arable lands are suitable for this crop. Recent developments promise a yield of 20 tons/hectare of fresh roots (ETA, 2009a). But, as a staple crop, it may not be the optimal choice for conversion into fuel ethanol.

Thus, the implementation of a program for the development and production of biofuels in Bénin will require the cultivation of thousands of hectares of energy crops; for this reason the identification of an optimal agricultural development model will be necessary, in order to ensure reliable and sustainable production of feedstock. With regard to this issue several options could be adopted, each one having its advantages and disadvantages, as summarized in Table 10.2. An integrated approach to these three models (contract farming, village-scale community-managed plantations and agro-industrial complexes)

TABLE 10.2 Feedstock Agricultural Development Models for Biofuels in Benin

Item	Contract Farming	Village-Scale Community Management	Agro-industrial Complex
Crop management	Benin cotton experience	Single-minded focus, management control	Complex multicrop management
	Farmers likely to get higher share of value added	Farmers likely to get less value-added	Farmers' share of value-added varies with crop
Technology change	Via experience and training	Higher rate of technology change likely	Rate of technology change varies with crop
Risk to steady supply	Higher	Medium	Lower
Risk to food supply	Higher	Medium	Lower
Overall farmers' development	Higher	Lower	Medium

would help to establish a balance between the need to achieve a high production capacity of cost-competitive, high-quality standardized biofuels (mainly through the adoption of the complex model, at least at the beginning), and the need to promote rural development, farmers' entrepreneurship and organizational and productive capacity (mainly through contract farming and village-scale plantations).

FINAL REMARKS

Lessons learned from the efforts in Kenya and Bénin to promote biofuels on a national scale could be useful to other African countries. Kenya's approach has relied mostly on private initiative backed by state support in various forms, but generally lacking in systematic government policy, legislation and regulation. No particular attention has been given to the management of technology change. Bénin has opted to launch its biofuels with a strong role for the state as legislator and regulator, attempting to implement policies that had in principle been vetted by relevant stakeholders in the country. However, like Kenya, Bénin appears not explicitly concerned with technology change and technology absorption, except for its emphasis on small-scale family agriculture as a source of biofuel feedstock. In neither country has a national biofuel program been firmly established today. Ethanol is produced in both countries, but the markets supplied are the traditional potable, pharmaceutical and industrial.

With respect to establishing national biofuels economies in any country, but especially African countries, the Kenya and Bénin and other countries' experiences suggest the following recommendations:

Improve government and institutional support to provide an enabling environment, but not necessarily control

Bénin is trying to set up the enabling environment, but should consider the extent of government involvement and its effect on the ability to attract private investment. Kenya, on the other hand, has been able to launch projects but has failed to create a national program for lack of adequate institutional support and coordination, and stakeholders' consensus. Other countries in Africa should consider the appropriate balance of private and government engagement to ensure the sustainable implementation of their national biofuel programs (Mckenzie-Hedger, 2000).

Develop clear, consistent, sustainable policies

As a late entrant to the biofuel economy, Bénin is pursuing initiatives that might lead to consistent and sustainable policies, although it may wish to review the need for appropriate balance between government and private initiatives. Kenya, an early entrant into the biofuel economy, has run through the early stages of its learning curve, but still has work to do to achieve sustainable policies.

Provide capital and pricing incentives

In the history of national biofuels programs, there is no example of success without an initial government push that reduces the risk of private capital investment in the biofuel market. This is especially relevant at times, such as presently, when crude prices are low and likely to remain low for a while. However, although such incentives are crucial for the launching stage, they should be lifted gradually over time, lest the will to innovate and improve competitiveness is dampened.

Promote stakeholders' dialogues and transparent public–private partnership

As Bénin wisely understood, the engagement of the relevant stakeholders in the design and implementation of national biofuel programs is a necessary condition for their sustainability. Such dialogues cover all kinds of concerns, including those about technology change and absorption, allowing them to be expressed, considered and acted upon.

Build up local construction and technical and managerial capacity

Although at first foreign designs, project management and know-how may be required to launch a national program, the quicker local construction, technical and managerial competences are developed, the better the chances of achieving sustainability. This is why it is important to factor in the technology transfer and absorption components of a national biofuel program early on (Trindade, 1994, 2000, 2007).

Encourage simpler designs and staged expansion

To facilitate the implementation of projects and the first steps in the technological learning curve, consideration should be given to the simplest, flexible and economic process designs available for biofuel production. This approach also facilitates the engagement of local equipment manufacturers, the training of operators and the management of technology change and absorption. As the program evolves, the successive capacity expansions can consider adopting more advanced economic technologies.

Make sure sustainable feedstock cost and supply are maintained (Trindade and Yang, 1982)

The single most important cost item in the final cost of biofuels is the cost of feedstock, which can range between 50% and 80% of total cost. Therefore, the economic feasibility of biofuels depends heavily on the reliable, flexible and economic supply of feedstock. The capital costs of equipment idle due to lack of feedstock supply can easily kill off even the best project.

Develop strategies to meet feedstock shortages

The reliable supply of economically priced feedstock is so crucial that one useful strategy to consider, as exemplified by the case of Bénin, is the utilization of flexible multiple feedstocks, such as sugarcane and sweet sorghum or cassava.

Future of bioethanol in Africa

Despite the large potential for bioethanol in Africa and the starts and stops of many African countries, the current use of fuel ethanol seems to be very limited in that continent. With respect to other countries in Africa, the prospects of national bioethanol programs hinge on contradictory forces, namely, the relatively low costs of crude in the short term and the pressures to combat global warming by decreasing greenhouse gas emissions in the long term. African countries embarking on bioethanol initiatives will not be many and most likely will be those which have sufficient land, water, human skills and access to financing to produce sufficient food, and where the relevant stakeholders can agree on a flexible but coherent set of policies and incentives to move forward on the bioethanol pathway.

Biofuels experience in other continents could help Africa as well as international cooperation with other countries (Government of Brazil, 2010; UK in Brazil, 2006). As the African experience has shown so far, the road to a national biofuels economy is not a straight line. This is a stepwise process with steps forward and backward. Such processes take time, but can be accelerated by strong stakeholders' consensus at national level, higher oil prices, mobility technology changes and the pressure to curb greenhouse gas emissions.

REFERENCES

Agra,   www.agra-net.net/agra/world-ethanol-and-biofuels-report/biofuel-news/ethanol/ethiopia-fuelethanol-use-crosses-50-mln-litre-threshold-1.htm.

Anonymous, 2010. Kenya Targets E10 Blend.

Baraka, M., 1989. Alternative Energy for Transport Industry – The Kenyan Experience with Ethanol. Internal Document, Sales and Technical Services, Caltex Oil (Kenya) Limited.

Batidzirai, B., 2007. Bioethanol Technologies in Africa. In: UNIDO/AU/Brazil First High-Level Biofuels Seminar in Africa, Addis Ababa, 30 July–1 August 2007.

Business Daily Africa, http://www.businessdailyafrica.com/-/539552/679148/-/59flfw/-/index.html.

CNBCAFRICA.com, 2015. Kenya turns to privatisation of sugar mills. http://www.cnbcafrica.com/news/east-africa/2015/05/18/kenya-privatisation-sugar-mills/ (18.05.15.).

Crilly, R., 2008. Kenya taps into Brazil's ethanol expertise. www.csmon-itor.com/World/Africa/2008/1114/p11s01-woaf.html?sms_ss=buzz.

Econergy International Corporation, 2008. Mozambique Biofuels Assessment, Final Report, May 1st.

Endevelu/ESD (Endelevu Energy and Energy for Sustainable Development Africa), 2008. A Roadmap for Biofuels in Kenya – Opportunities and Obstacles. A Report to GTZ/Kenya Ministry of Agriculture.

ETA, 2009a. Étude de Faisabilité de la Production de Biocombustibles Modernes (Bioethanol et Biodiesel) au Bénin et Élaboration de la Stratégie de leur Promotion – Rapport Final de la Phase 1 – Analyse de la Faisabilité Technique, Financière, Économique, Insti- tutionelle et Environnmentale, Cotonou, février, ETA, Florence.

ETA, 2009b. Stratégie pour la Promotion des Biocarburants au Bénin, Cotonou. ETA, Florence.

Ethco, Ethanol: Malawi's Versatile Fuel, one page brochure, and www.ethanolmw.com.

Government of Brazil, 2010. Declaração Conjunta Brasil–Índia – Brasília, 15 de abril de 2010 (Joint Declaration of Brazil–India, Brasilia, 15 April 2010). www.itamaraty.gov.br/sala-de-imprensa/notas-a-imprensa/2010/04/15/declaracao-conjunta-brasil-india-brasilia-15-de/?searchterm=Brasil-India Memorandum de Entendimento.

Leal, M.R.L.V., 2016. Personal communication.

Mckenzie-Hedger, M., 2000. Enabling environments for technology transfer. In: Metz, B., Davidson, O., Martens, J.-W., Van Rooijen, S., Van Wie Mcgrory, L. (Eds.), Methodological and Technological Issues in Technology Transfer Intergovernmental Panel on Climate Change, Cambridge University Press, pp. 105–141.

Mozambique,   www.treehugger.com/renewable-energy/worlds-first-ethanol-cooking-fuel-plant-opens-mozambique.html.

Nigerian National Petroleum Corporation Researchomatic, 2013. Retrieved 2, 2013, from http://www.researchomatic.com/Nigerian-National-Petroleum-Corporation-154020.html

Nigeria, www.nairaland.com/56357/ethanol-fuel-alternative-fossil-fuel.

Odhiambo, A., 2009. Mumias eyes new revenue stream from ethanol plant. Business Daily http://www.businessdailyafrica.com/-/539552/823086/-/69e6tt/-/.

Scurlock, J., Rosenschein, A., Hall, O.D., 1991. Fuelling the Future: Power Alcohol in Zimbabwe. African Centre for Technology Studies (ACTS), Nairobi.

Senegal, http://ejatlas.org/conflict/sen-huile-sen-ethanol-biofuels-senegal.

South Africa, 2015. http://www.ethanolproducer.com/articles/10329/south-africa-to-mandate-biofuels-blending-startingin-2015.

South Africa, 2015b. http://naija247news.com/2015/08/cheaper-oil-forces-south-africa-to-rework-biofuels-subsidy/.

Trindade, S.C., 1982. Review of Power Alcohol Issues in Kenya. Report on Integration of Power Alcohol Operations with the UN/TCD Project KEN/76/005. Energy Planning, Policy and Programming in Kenya.

Trindade, S.C., 1994. Transfer of clean(er) technologies to developing countries. In: Ayres, R.U., Simonis, U.E. (Eds.), Industrial Metabolism: Restructuring for Sustainable Development United Nations University Press, New York, pp. 319–336.

Trindade, S.C., 2000. Managing technological change in support of the climate change convention: a framework for decision-making. In: Metz, B., Davidson, O., Martens, J.-W., Van Rooijen, S., Van Wie Mcgrory, L. (Eds.), *Transfer*, Intergovernmental Panel on Climate Change Cambridge University Press, pp. 47–66.

Trindade, S.C., 2007. Transfer of technology and expertise. In: Worldwatch Instititute, Biofuels for Transport: Global Potential and Implications for Sustainable Energy and Agriculture Earthscan, London, pp. 263–275.

Trindade, S.C. and Yang, V., 1982. Resources and resource conversion. In: Proceedings of the Fifth International Alcohol Fuel Technology Symposium, 4, Auckland, New Zealand, pp. 173–192.

Trindade, S.C., et al., 2012. Biofuels technology change management and implementation strategies: lessons from Kenya and Bénin. In: Johnson, F.X., Vikram, S. (Eds.), Bioenergy for Sustainable Development and International Competitiveness: The Role of Sugar Cane in Africa. Routledge of Taylor & Francis Group, London, pp. 369–389.

UK in Brazil, 2006. Joint Statement by President Luiz Inácio Lula da Silva and Prime Minister Tony Blair, 9 March 2006, http://ukinbrazil.fco.gov.uk/en/news/?view=News&id=2056244.

UN, 1991. Energy systems, environment and development – a reader. Advanced Technology Assessment System, 6, New York, Autumn, p. xvi.

Wosemo, 2015. Council grants Kenya sugar safeguard. http://www.comesa.int/summit2015/council-grants-kenya-sugar-safeguard/ (26.03.15.).

Zambia, http://www.biofuelsdigest.com/bdigest/tag/zambia/.

Zimbabwe, 2014. www.dailynews.co.zw/articles/2014/01/14/zim-faces-fuel-crisis.

Chapter 11

Future of Global Bioethanol: An Appraisal of Results, Risk and Uncertainties

P. Lemos[1] and F.C. Mesquita[2]

[1]Laboratory for Studies on the Organization of Research & Innovation, Department of Science and Technology Policy, University of Campinas, Campinas, Brazil [2]Department of Science and Technology Policy, University of Campinas, Campinas, Brazil

INTRODUCTION

This chapter aims to analyze certainties and uncertainties related to the future of bioethanol as a global commodity. On the one hand, bioethanol has shown a trend of growth in the past decade at least in two main countries: the US and Brazil. New technologies (as the so-called second generation, based on cellulosic raw material) are in their first steps to get technical and commercial feasibility and still are not guaranteed. On the other hand, biofuel consumption is growing only in a limited number of countries of North and South America. Markets in Europe and Asia are still low and uncertain.

One of the main challenges for bioethanol growth is to overcome its market limitations. It brings back the question about the economic feasibility of transforming bioethanol into a global commodity. After sparking a debate at the end of the 2000s, the real productive, technological and diplomatic efforts to complete this process seem to have lost momentum. Is it possible to think in new directions related to this process in the coming years? In order to achieve a global scale in international trade, bioethanol has to be not just a complementary blend with gasoline but a competitive fuel against other renewable sources and technologies such as electric engines, for example. At the same time that new renewable energy sources are becoming effective options, primary sources like coal and oil remain growing. Changes in the global energy matrix are showing a very competitive race between renewable energy and fossil fuels. More than this, the competition is also within these categories: different types of biofuels compete among themselves (eg, bioethanol × biodiesel) and the same happens to fossil fuels (oil × shale gas).

The chapter presents four subitems after a brief description of the main factors and patterns behind the potential process of global commoditization of bioethanol. The first item emphasizes data analysis of bioethanol production and consumption and its impacts on the international market. The second presents a comparative analysis of bioethanol against traditional and some alternative fuels, technologies and products. The third discusses mainly the projections and main challenges of bioethanol to become a truly global energy commodity. Finally, we present our final remarks about the futures of bioethanol worldwide.

INTERNATIONAL MARKETS OF BIOETHANOL: PRODUCTION AND CONSUMPTION

Doubts about the potential growth of bioethanol are not recent. At least since the oil crises in 1973, when bioethanol emerged as an alternative fuel, several studies have been carried out to analyze the possibility of sustainable growth for this renewable fuel. At that time, different challenges were pointed out. Overcoming technical barriers in the automobile industry, not interfering with fuel prices, and creating a market both as fuel and a chemical product, are some examples (Goldemberg, 1979; Gao, 1980; von Bremen and Schmoltzi, 1986).

Despite bioethanol market still being uncertain, the economic, technological and political context has changed for renewable energies (Jacobsson and Johnson, 2000). Bioethanol is a reality in a few countries. The increasing concerns of environmental problems (eg, with the air quality in the metropolises), brought bioethanol again to the core of debates about alternative fuels (Goldemberg et al., 2008). Therefore, besides being an alternative for reducing oil dependence in the context of energy security policies, bioethanol has also gain a sustainability attribute when it was considered one of the solutions to deal with the reduction of CO_2 emissions (Mussatto et al., 2010).

Some of the limitations bioethanol faced in the 1970s and the 1980s were overcome (Walls et al., 2011; Searle et al., 2014). Nowadays it can be employed both as a pure fuel or in combination with fossil fuels. On the other hand, agricultural development, mainly in corn and sugarcane crops, reduced the bioethanol production costs in the first-generation process (Hettinga et al., 2009). It is also interesting to mention that promoting bioethanol is also used as a strategy for support rural communities, in the context of agricultural and social policies.

The world's bioethanol production passed from 65,000 barrels per day in 1980 to 1,471,000 in 2012 (EIA, 2015). Nevertheless, the pace of growth is quite different among countries. Demand for bioethanol has boosted in the United States and Brazil after the 2000s and, with a smaller force, in the European Union (EU). Countries in Asia have less clear directives, but have also stimulated bioethanol production and consumption.

As discussed in this book, bioethanol production and consumption, regardless of the feedstock employed, are a matter of energy policy and consequently are strongly biased by national strategies. Its growth depends on several variables such as oil availability and prices, government efforts, agricultural development, food policies and on other alternatives that compete with bioethanol. Table 11.1 presents how production and consumption of bioethanol increased worldwide between 2005 and 2012.

As can be seen in Table 11.1, North America (particularly the US) led the growth observed in that period. In 2003, the US became the world's largest consumer of bioethanol. In 2006 that country overcame Brazil, as presented in Fig. 11.1, also becoming the largest producer.

The sharp increase in bioethanol production in the US started in 2002. During the 10 years between 2002 and 2012, the number of bioethanol plants went from 61 to 209 (RFA, 2015). Most of these are located in the American Mid-West, being part of the traditional agribusiness complex based on corn production.

The main incentive for bioethanol in the US is the domestic market. There are governmental incentives to blend bioethanol with gasoline (Walls et al., 2011) not only for low percentages of ethanol, but also for the so-called E85 which refers to engines using a blend of 85% of bioethanol and 15% of gasoline. From 2003 to 2010 the number of these vehicles rose from 179,090 to 618,505 (EIA, 2012). Compared with other alternative vehicles, E85 is the one that is growing faster. However, the distribution structure of bioethanol stations is concentrated in the producer states, like Illinois and Iowa.

South America is the world's second largest bioethanol producer and consumer; mostly because Brazil concentrates more than 90% of the regional production. In this country, significant investments on bioethanol date from 1975, with the National Alcohol Program (Proalcool). Based on sugarcane processing, Brazil soon became the world's largest producer. The sector had difficulties in the beginning of the 1990s, when national policies toward biofuels were left behind (Furtado et al., 2011). The production rose again after 2003 induced by the introduction, diffusion and sale of flex-fuel vehicles. In 2014, 88.2% of domestic vehicles used flex-fuel motors (Anfavea, 2015). Nevertheless, since 2008 Brazilian bioethanol consumption has presented unstable growth (Fig. 11.1). Uncertainties regarding public policies towards biofuels' relative prices have raised doubts about the future of this industry (Salles-Filho et al., 2015).

Besides Brazil, bioethanol is also becoming an important sector in Colombia and Argentina. Colombia has a national biofuel program that, since 2005, has incentivized a blended of 5% of ethanol in gasoline. Projections about bioethanol growth in the country are optimistic (Valencia and Cardona, 2014). Argentina has more expertise in producing biodiesel from soybean, wherein since 2006 this has been produced on a large scale. Bioethanol, on the other hand, is a recent industry. Excluding the failed experience between 1984 and 1988, bioethanol, made from corn and sugarcane, started as a new industry only after 2010 (Timilsina et al., 2013). Recently, the Argentinian government has raised the share of bioethanol in gasoline to 10%.

TABLE 11.1 Fuel Bioethanol Production (P) and Consumption (C) From 2005 to 2012, in Thousand Barrels per Day

Year	North America		Central and South America		Europe		Eurasia and Middle East		Asia, Oceania and Africa	
	P	C	P	C	P	C	P	C	P	C
2005	259.1	270.6	284.7	184.0	14.8	19.7	0.3	0.0	26.2	26.2
2006	323.0	362.1	328.3	200.2	27.3	31.2	0.5	0.2	36.8	35.8
2007	439.2	469.2	414.6	268.2	31.4	40.0	0.7	0.4	38.6	38.5
2008	620.6	653.9	497.8	342.8	47.4	61.6	0.7	0.6	48.8	48.7
2009	733.5	745.9	476.5	402.9	59.3	77.7	1.3	0.6	52.5	53.6
2010	891.7	871.1	502.9	393.5	71.6	98.4	1.2	0.7	53.5	55.5
2011	938.9	883.4	415.9	350.1	72.8	104.3	0.4	0.6	62.5	68.7
2012	908.3	879.5	428.9	378.9	68.5	114.1	0.3	0.6	64.2	73.9

Source: from EIA (2015).

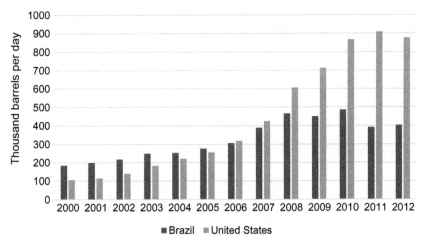

FIGURE 11.1   Bioethanol production in Brazil and the US between 2000 and 2012. *From EIA (2015).*

In Europe, although the use of bioethanol is still modest compared with North and South America, it is gradually changing in some countries. Considering the data from 2012, Germany is mainly responsible for the increase in the consumption of bioethanol. This country, a forerunner in the European Union to promote the use of biofuels (mainly biodiesel) (Kaup and Selbmann, 2013), established in 2009 a mandatory share of 2.8% of bioethanol in gasoline (Sorda et al., 2010). The second major consumer in Europe is the United Kingdom. Policies to stimulate bioethanol blend in gasoline in this country started in 2005 with an initial projection to reach 5.75% in 2010 (Bomb et al., 2007). Nowadays, petrol fuels containing 5% of bioethanol (E5) are widespread in the country and there are plans to improve the option for a 10% mixture (E10). France is the third major consumer. This country has a mandatory share of bioethanol in petrol fuels that passed from 1.2% in 2005 to 7% in 2010 (Sorda et al., 2010), a value that is the same in 2015. It is important to mention that Europe is becoming an important market for bioethanol thanks mainly to the EU directives towards biofuel consumption and production.

Nevertheless, other parts of the world bring less optimistic perspectives for bioethanol growth. In Eurasia and the Middle East, both consumption and production are nearly zero. In Asia, some countries are creating policies in favor of renewable fuels, but it is still unstable. In India, the second largest producer of sugarcane after Brazil, bioethanol has been produced from sugar molasses, a residue of sugar processing. Recently, the government established a mandatory blend of 10% of ethanol in petrol fuels for 2016 (Economic Times, 2015). In China, as we can see in this book, bioethanol production and consumption are facing an unstable scenario: firstly, because of the feedstock, mostly based on grains that compete with food supply (Lammers et al., 2014); secondly, because the national energy policy is somehow ambiguous when referring to fuel consumption. In Africa, bioethanol growth is dealing with the social challenge of creating jobs and not affecting food prices. The production is on a small scale and based on sugarcane, corn and cassava. The industry is more consolidated in South Africa. Countries like Zimbabwe and Kenya have also had experiences in this area. Nevertheless, there was no continuity to these programs (Bensah et al., 2015).

BIOETHANOL SUPPLY AND DEMAND: PROS AND CONS

As a new technological trajectory (compared to prevalent trajectory based on fossil fuels), the wide diffusion of renewable fuels is expected to be a slow, painful and uncertain process (Jacobsson and Johnson, 2000). The path of bioethanol growth has obstacles and opportunities that we try to deploy in the following sections.

Cons

Bioethanol growth depends directly on the developments made in the automobile industry. The electric car is probably the one that deserves most attention. This technology has a long trajectory and it is now becoming more and more attractive (Høyer, 2008).

Like biofuels, electrical vehicles had a greater projection after the 1980s, when the environmental concerns became stronger in a global perspective. At the same time, some technical and economic barriers to create hybrid electric vehicles (HEV) were overcome (Høyer, 2008; Dijk and Yarime, 2010). A fueling infrastructure for electric vehicles has been created mainly in Europe. Countries like Germany, France, the UK and Denmark, for instance, are directing efforts to this sector, making HEV worthwhile for consumers (Dijk et al., 2013). In the US it is important to highlight the experience of the city of Los Angeles (Kim and Rahimi, 2014).

In conjunction with new alternative engines in the automobile industry, the challenge for bioethanol growth depends on the competition with fossil fuels, especially gasoline. As is presented in Table 11.2, the consumption of gasoline has shown different patterns in different countries.

Considering the growth rates between 2005 and 2012, the region where gasoline consumption increased fastest was Central and South America. Paradoxically, this is one of the places where bioethanol policies are more consolidated, showing that even in the productive regions the competition with gasoline is far from being over. Asia (boosted mainly by China) also has shown high growth rates of gasoline consumption. Africa, Eurasia, and the Middle East follow a similar condition. The reduction is modest (but important) in North America. Gasoline production and retail continue to be one of the most competitive sectors of the US economy (Holmes et al., 2013). Only Europe has been facing a relevant decrease in gasoline consumption, essentially because in the

TABLE 11.2 Consumption of Gasoline From 2005 to 2012, in Thousand Barrels per Day

Regions	2005	2006	2007	2008	2009	2010	2011	2012
North America	10,522.7	10,662.7	10,758.7	10,483.4	10,502.5	10,539.9	10,304.7	10,251.8
Central and South America	1,071.2	1,097.5	1,171.2	1,151.8	1,161.5	1,246.5	1,363.7	1,502.7
Europe	2,782.7	2,694.8	2,603.6	2,483.2	2,410.1	2,296.2	2,219.3	2,093.7
Eurasia and Middle East	2,143.3	2,310.0	2,265.3	2,360.0	2,384.1	2,457.9	2,544.7	2,603.5
Asia, Oceania and Africa	4,652.9	4,718.5	4,901.0	4,947.7	5,281.0	5,620.5	5,786.1	6,144.9
World	21,172.8	21,483.4	21,699.7	21,426.0	21,739.2	22,161.1	22,218.4	22,596.5

Source: from EIA (2015).

EU diesel is spurred by favorable excise taxes. Actually, Europe is showing a decrease in the total consumption of oil products, particularly since 2008, when the financial crisis helped change some consumption patterns in that region. In this period, the overall demand for oil products has declined by 8%, particularly due to the decline of gasoline (FuelsEurope, 2014).

The EU directives of 2009 pointed to a minimal participation of 10% of renewable fuels in transportation by 2020. New directives are now being discussed in the EU Parliament as a consequence of the 21st Conference of the Parties held in Paris in 2015. Among these it is worth mentioning the possibility of introducing a minimal participation of the so-called "advanced biofuels" (as seaweeds or cellulosic raw material that do not compete with food production in land use).

Biodiesel in Europe is so far the main biofuel being boosted by governmental policies. The EU capacity to produce biodiesel in 2014 was approximately of 23 million tons (EBB, 2015) and is growing. The perspectives are to increase the production and consumption of mixtures of 10% and 20% (B10 and B20) and eventually B100. Ethanol in Europe is a secondary biofuel, but still shows some potential. This theme is better developed in other chapters of this book.

Pros

The opportunities for biofuel growth are both in the supply and demand side. In production the change that has brought most enthusiasm to the sector is second-generation bioethanol. This technology consists of the use of enzymes, in a biochemical process, to break cellulose molecules into thousands of simple sugars that can be fermented to produce bioethanol (Raele et al., 2014). There is also the possibility of breaking cellulose feedstock in a thermochemical conversion. In this system, the feedstock is gasified to produce synthesis gas $(CO + H_2)$ that is fermented or catalytically converted to ethanol (Dwivedi et al., 2009; Sims et al., 2010).

Second-generation technology created the possibility of producing bioethanol from different types of biomass, like corn stover, sugarcane bagasse or wood, amongst others. Recent improvements in pretreatment techniques of the feedstock and the development of enzymes promise to increase the efficiency of the second-generation process (Sims et al., 2010). It is expected that in addition to the increasing productive capacity in traditional regions of the US and Brazil, this technology could stimulate the growth of bioethanol in new areas of expansion like Africa for example (Bensah et al., 2015).[1]

New projections are showing the potential of third-generation bioethanol produced by marine biomass (Alaswad et al., 2015). Some of the social and economic advantages are that algae biofuels does not compete with agriculture,

---

1. See Chapter 10 in this book about bioethanol in Africa.

having lower impact in terms of land use changes (LUC). Nevertheless, it also should be considered that the productive process creates high rates of ammonium concentration, which can be damage for local environment (Alaswad et al., 2015). Algae can also contribute to the production of different types of biofuels, such as for biodiesel, biogas or bioethanol (Andersson et al., 2014).

Regarding the demand side, besides the changes mentioned above in national policies which are spreading the consumption, bioethanol is becoming a viable alternative for the airline industry (Cremonez et al., 2015). It has the potential of connecting the product with an industry that has optimistic projections of growth in an economy increasingly depending on global communication. In Brazil, light aircraft, especially in agricultural aviation, are already using bioethanol. New tests are now being carried out in this country revealing quite promising results for heavier aircrafts. In this area, bioethanol caused both a decrease in operating costs and an environmental improvement (Cremonez et al., 2015). However, considering the global scale of aviation bioethanol is not yet an option (Nygren et al., 2009). It can become a real alternative for aviation in the moment its supply becomes available everywhere.

BIOETHANOL: A TRULY GLOBALLY TRADED ENERGY COMMODITY?

The main goal of this section is to discuss the projections and scenarios of the global bioethanol markets, in the context of the evolution and development of a commoditization process.

To meet this goal it is necessary to analyze a set of questions about biofuel market creation and development, the present conditions of diffusion of bioethanol worldwide. Foreign bioethanol trade is an obvious element for understanding this process, along with considerations about production and consumption on a global scale.

The methodology here employed is based on the analysis of trends of production, consumption and international trade. For this purpose we use two main sources of data and information: FAO/OECD Agricultural Outlook (OECD/FAO, 2015) and a comprehensive analysis focused on international trade (Junginger et al., 2013). Complementary references are also employed to highlight specific points about the globalization of bioethanol.

The 21st edition of the Agricultural Outlook, the 11th prepared jointly by FAO and OECD, presents projections to 2024 for major agricultural commodities in which biofuels are considered. The market projections cover OECD member countries (European Union as a region) and FAO member countries such as Brazil, the Russian Federation, India, China and South Africa. The other publication aforementioned (Junginger et al., 2013) brings a global perspective of biofuels and biomass as commodities traded worldwide in significant volumes and values. This publication is a synthesis of the "Task 40," an international network and working group acting since 2004 under the IEA (International Energy Agency) Bioenergy Implementing Agreement to investigate biomass production and supply.

The Importance of Bioethanol Policies: Blending, Targeting and Potential (Dis)continuities

One of the main factors that act in favor of bioethanol production, consumption and international trade is the environmental regulatory framework adopted in different countries and regions. Policies in general are based on drivers of bioethanol blend mandates (which can have different recommendations of mixtures from E5 to E100), subsidies (mainly taxes exemption) and certification schemes (Van Meijl et al., 2015). After the conclusions of the 21st Conference of the Parties on Climate Change in December of 2016 these policies will probably be based on more restrictive levels of emissions and are more favorable to biofuel immediate production.

One of the consequences of all these initiatives is the development of bioenergy solutions that reached an exponential growth in some countries, raising bioethanol in the Americas (and biodiesel in Europe) to volumes and amounts typically achieved by commodities. Nevertheless, instruments planned in order to protect domestic markets have also generated distortions and frictions at the international markets such as for tariffs, quotas, subsidies to domestic producers and hard-to-reach technical exigencies.

According to the FAO/OECD Outlook global ethanol production is projected to increase during 2014–2024 from about 114 to 134.5 billion of liters. However, this increase is mainly due to the supply of domestic markets. The US, Brazil and EU remain the most important producers followed by the People's Republic of China. Fig. 11.2 shows a forecast of bioethanol production in the next 10 years by countries and regions (OECD/FAO, 2015).

A similar trend can be verified on the stability of bioethanol international trade growth during the period covered by the FAO-OECD Outlook. According to this report neither the United States nor Brazil will dramatically change their positions in the international markets by 2024.

The supply of sugarcane-based ethanol is expected to be limited and decreasing over time, along with very little growth of cellulosic ethanol. Exportations of sugarcane bioethanol from Brazil are expected to rise from 2.2 bilion liters (average 2012–2014) to 3.5 billion liters by 2024. This is a quite frustrating scenario if compared to the expectations presented by some authors some years ago (Cerqueira Leite et al., 2007).

With respect to other developing countries, the scenario forecasts new net exporters of ethanol such as Argentina reaching exports of 0.6 billion liters by 2024. The European Union, Japan and Canada will maintain their importance of major ethanol importers expanding circa 1.1 billion liters during the forecasted period.

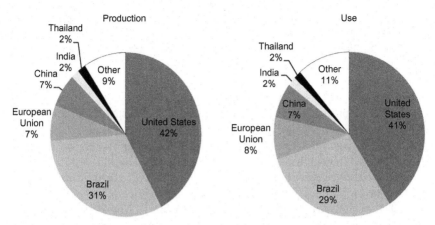

FIGURE 11.2   Global forecast bioethanol production and use in 2024. *From OECD/FAO, 2015. OECD-FAO Agricultural Outlook. OECD Agriculture Statistics (Database),* http://dx.doi.org/10.1787/arg-outl-data-en.

Notwithstanding, the major uncertainty raised by OECD/FAO analysis refers to national policies towards bioethanol. There is strong uncertainty about how countries will shape their internal energy policies. The goals of replacing fossil fuels by biofuels and the investments in different types of renewable are ambiguous both in developed and in emerging economies. The uncertainty increases as the new agreement on climate change issued in December of 2016 starts to be adopted by the parties.

As for emerging economies, the OECD/FAO Outlook remains with the same low in optimism vision. It argues that audacious biofuel production targets envisaged in recent years have been reviewed in order to accommodate changes in domestic sectors and a possible reduction in exportation opportunities. However, the biofuel industries in major producing countries could counterbalance this trend in the short term, turning to the domestic market. In Brazil, for instance, recent policy instruments based on taxation and blending requirements can function as a local incentive to the bioethanol sector. Flex-fuel vehicles will remain with an important role over the next decade. Additionally, the effects of the exchange rate policy based on devaluation of Brazilian currency (Real) facing the US dollar and the recovery of domestic prices from gasoline can influence a different performance of Brazilian production and foreign trade for bioethanol.

Global Governance: The Process of Standardization and Harmonization in Technical and Sustainability Matters

Another important challenge for transforming bioethanol in a global commodity is to evaluate the efforts of global governance. It is essential in order to coordinate and articulate policy, economic and technological differences between countries (Lima and Gupta, 2013).

This coordination has much to do with technical (and nontechnical) trade barriers. For instance, corrosion inhibitors, chemical uniformity properties, quality standards, market uniformity for different labels, energy efficiency to mention a few, are tasks that need huge efforts of coordination.

The development of worldwide harmonized technical standards for bioethanol needs to reproduce global instruments and institutions already existing in national contexts. International regulatory and metrology boards are particularly needed to accomplish this endeavor (Brandi et al., 2011).

The bilateral relationship between the US and Brazil could be an example to accomplish technical standards and harmonization in a supranational scale. In 2007 these countries signed a broad Memorandum of Understanding (MoU), specifically addressing biofuels cooperation, with aspects on technology sharing between the countries, feasibility studies and technical assistance to build domestic biofuel industries in third countries, and multilateral efforts for the global development of biofuels.

One attempt to strengthen relationships in a tripartite dimension including Brazil, the US and the EU was carried in December 2007 with the publication of the "White Paper on Internationally Compatible Biofuel Standards," a document presenting the framework for further development of measurement standards and for documentary standards, two important elements of the standardization process. The most important result of the tripartite efforts was the development of certified reference materials (CRM), devices and certificates for stable, homogeneous and highly reliable substances, property and uncertainty measurement values, applied in equipment calibration, quality control and method validation (Brandi et al., 2011).

Regarding sustainability issues there are plenty of technical, conceptual, methodological (and probably, ideological) convergences and divergences towards harmonization, standardization and orchestration governance (Faaij et al., 2014). As verified in other sectors and markets, standardization is crucial to guarantee transparency in markets, helping to remove obstacles to production and international trade (Pacini, 2015).

The international statistical system for bioethanol also suffers from an absence of harmonization and standardization schemes. Available statistics of volume and prices of international trade are not totally reliable and it is far from other energy databases such as for oil and gas.

According to Lammers et al. (2014), trade statistics related to bioethanol are imprecise because the various potential end uses of ethanol like fuel, industrial processes and beverage production, and the lack of adequate classification codes for biofuels, normally codified by countries using specific classification. In this sense, foreign trade of fuel ethanol has a very low capability to distinguish from trade in denatured, nondenatured and mixed ethanol blends because of the absence of a harmonized system base for classification in terms of end-use. Therefore, this very undeveloped system of statistics can be considered a barrier to the expansion of bioethanol trade in a worldwide dimension.

The Different Approaches of Leading Countries for Global Commoditization Process

According to Roehrkasten (2015) the globalization of bioethanol seems to be a goal spurred by the Brazilian diplomacy but not followed by other countries— at least not with the same enthusiasm. The perspectives of having a cluster of countries engaged in promoting bioethanol as a global commodity bumps into political obstacles that cannot be easily overcome (Lammers et al., 2014). Even if it one takes into consideration the fact of having common interest, the two leading producers and exporters of bioethanol in the world (Brazil and the US) have quite different strategies regarding energy and biofuels.[2]

Since 2010 the exports of US ethanol has been twice the Brazilian exports. In 2012, Brazil and the US supplied ethanol for about 50 countries each, with 19 common countries around the world (International Energy Agency; Brazilian National Petroleum Agency).

Nevertheless, the development of international trade does not seem to be a strategic goal for the US biofuel industry. The US bioethanol industry is much more oriented to achieve domestic markets and accomplish with regulatory goals than to develop a global market for biofuels. That does not mean this industry will not undertake strategies of expansion in the future. For instance, the Renewables Fuel Association (RFA), the most important industrial institution representing bioethanol sector and its value chain criticizes the so-called "blend wall" restriction saying that this is part of a misinformation war architected by the oil industry. The term "blend wall" can be understood as the limited percentage of bioethanol that can be blended into gasoline.[3] Apart from lobbying activities undertaken by the industry, there are no big corporative efforts to make bioethanol global. There is no robust evidence of joint collaboration and cooperation efforts perpetrated by the US and Brazil, since the initial phase of interactions in 2007 with the signature of the MoU aforementioned (Meyer, 2014).

In this context, it is difficult to sustain, at least so far, that the internationalization and globalization of the bioethanol market can be considered a collective goal that encompasses the leading countries. It is quite reasonable to assume, aligned with the OECD/FAO analysis, that bioethanol sectors in various

---

2. See in this book the excellent overview of Michael Griffin, Bradley Saville, and Heather MacLean about biofuels policies in the US (chapter 2). Also the chapter 4 by Luciano Sousa et al. discusses the differences between the systems of innovation of USA and Brazil.

3. Jointly with the RFA, see for example the efforts of the US industry like "American Coalition for Ethanol" and "The Blend Your Own (BYO) Ethanol" campaign, just to cite two initiatives aiming to promote bioethanol in the context of the economic, technological, and ecological race between bioethanol and gasoline.

countries remain very influenced by a combination of price trends and public policy initiatives that, in the end, makes production and consumption targets vary considerably across countries without unified targets collectively planned and executed (OECD/FAO, 2015, 126).

FINAL REMARKS

Bioethanol has demonstrated its value as an alternative fuel to gasoline, in fact, more as a complement than a competitor. This comes mainly from the national policies implemented by countries, the recent increase of oil prices, the assumed importance of climatic changes questions, and the expected production of ethanol from cellulosic and other second- or even third-generation sources.

Bioethanol can be considered a restricted international commodity because of its importance in a few countries. It seems reasonable to assume that the globalization of bioethanol as a worldwide commodity remains in an initial phase of national incentive policies with impacts on international trade. A second phase of commoditization is still far from being achieved. Farther still is a third phase of liberalization removing barriers of all kinds.

Despite the expectations for high bioenergy demand on a global scale, the considerable development of the international market, and the growth of production, consumption and international trade in the last decade, the process of commoditization of bioethanol remains within the status of *potential*, aligned with the same trend verified in the development of biofuel markets in general. The scenario produced by FAO/OECD for the next decade is clear to demonstrate the modest performance expected for the bioethanol in terms of productive growth and foreign trade flows and volumes.

As for the policy contexts it is important to consider how the targets of biofuel consumption adopted by various countries will be continued, extended, reduced or terminated. It is necessary to note the movements of the most important markets for biofuels. In the US there is an absence of a final rulemaking by the United States Environmental Protection Agency (EPA) for the time being. The European Union's 2030 Framework for Climate and Energy Policies adopted in October 2014 did not define clear targets in general for biofuels beyond 2020 (OECD/FAO, 2015).

Revision of policies can also be influenced by food prices, GHG emissions and environmental impacts, making countries reformulate biofuel blending targets, like Germany that revised its blending target for 2009 from 6.25 to 5.25% (Faaij et al., 2014). This absence of clear orientation about the targets related to a significant share of the worldwide market imposes more uncertainties against the permanence of this type of energy policy instrument, even considering the fact that—throughout the next decade—a number of emerging markets like India, Thailand, Colombia, Philippines, Vietnam and Nigeria are planning to increase their blending ratios (OECD/FAO, 2015). An additional source of uncertainty, probably more headed to boost biofuels, is the consequences of the recent Paris Agreement on Climate Change whose goals can deeply modify the current rules, policies and regulations.

Another major uncertainty linked with the policy hesitancy, is whether biofuels in general, and in particular bioethanol, will sustain a relatively privileged position in the context of broader energy policies. Nowadays the energy policies include a very diverse set of renewable products, services and technologies therefore not just restricted to biofuels for transport purposes. For example, in the EU, the 2009 Renewable Energy Directive (RED) foresaw that renewable fuels (liquids and nonliquids) should increase to 10% of total transportation fuel by 2020. But, according to the OECD/FAO scenario, the portion coming from biofuels will reach 7% by 2020 and additional gains related to the RED target should be influenced by the development of other energy sources and products like electric cars, for example.

Facts and performance evidences related to the growth of the production and trade markets in a worldwide dimension in the last decade are unequivocal. But the leap from the current significant levels in volumes and economic values that qualify bioethanol as a global commodity extensively traded worldwide cannot yet be verified.

Hence, apparently just new disruptive trends—or more-than-incremental ones—whether in policies, markets or technologies may change this context, reconfiguring the forecasted slow capacity of bioethanol to build a new trajectory of growth, globalization and commoditization.

REFERENCES

Alaswad, A., Dassisti, M., Prescott, T., Olabi, A.G., 2015. Technologies and developments of third generation biofuel production. Renew. Sust. Energ. Rev.s 51, 1446–1460.

Andersson, V., Broberg, S., Hackl, R., 2014. ScienceDirect Algae-based biofuel production as part of an industrial cluster. Biomass Bioenerg. 71, 113–124.

Anfavea (National association of Brazilian car builders), 2015. Annual statistics. Available from: <http://www.anfavea.com.br/tabelas.html>.

Bensah, E.C., Kemausuor, F., Miezah, K., Kádár, Z., Mensah, M., 2015. African perspective on cellulosic ethanol production. Renew. Sust. Energ. Rev. 49, 1–11.

Bomb, C., McCormick, K., Deurwaarder, E., Kåberger, T., 2007. Biofuels for transport in Europe: lessons from Germany and the UK. Energ. Policy 35 (4), 2256–2267.

Brandi, H.S., Daroda, R.J., Souza, T.L., 2011. Standardization: an important tool in transforming biofuels into a commodity. Clean Technol. Environ. Policy 13 (5), 647–649.

Cerqueira Leite, Rogério C., Leal, Manoel R.L.V., Cortez, Luis A.B., Griffin, Michael W., Scandiffio, Mirna G., 2007. Can Brazil replace 5% of the 2025 gasoline world demand with ethanol? Energy 34 (5), 655–661.

Cremonez, P.A., Feroldi, M., de Oliveira, C.D.J., Teleken, J.G., Alves, H.J., Sampaio, S.C., 2015. Environmental, economic and social impact of aviation biofuel production in Brazil. New Biotechnol. 32 (2), 263–271.

Dijk, M., Orsato, R.J., Kemp, R., 2013. The emergence of an electric mobility trajectory. Energ. Policy 52, 135–145.

Dijk, M., Yarime, M., 2010. The emergence of hybrid-electric cars: innovation path creation through co-evolution of supply and demand. Technol. Forecast. Soc. Change 77 (8), 1371–1390.

Dwivedi, P., Alavalapati, J.R.R., Lal, P., 2009. Cellulosic ethanol production in the United States: conversion technologies, current production status, economics, and emerging developments. Energ. Sust. Dev. 13 (3), 174–182.

EBB – European Biodiesel Board, 2015. <http://www.ebb-eu.org/stats.php>.

Economic Times, 2015. Government to make 10% ethanol blending mandatory from next year. Available from: <http://goo.gl/o1EIPn>.

EIA (Energy Information Administration), 2012. Anual energy review. Available from: <http://www.eia.gov/totalenergy/data/annual/showtext.cfm?t=ptb1005>.

EIA (Energy Information Administration), 2015. International Energy Statistics. Available from: <https://goo.gl/P4mTPR>.

Faaij, A., Junginger, M., Goh, C.S., 2014. A general introduction to international bioenergy trade International Bioenergy Trade. Springer, Netherlands.1–15

FuelsEurope. Dataroom. 2014. <https://www.fuelseurope.eu/dataroom>.

Furtado, A.T., Scandiffio, M.I.G., Cortez, L.A.B., 2011. The Brazilian sugarcane innovation system. Energ. Policy 39 (1), 156–166.

GAO, 1980. Potential of Ethanol as a Motor Vehicle Fuel. Washington. Available from: <http://www.gao.gov/products/EMD-80-73>.

Goldemberg, J., 1979. Renewable energy sources: the case of brazil. Nat. Res. Forum 3, 253–262.

Goldemberg, J., Coelho, S.T., Guardabassi, P., 2008. The sustainability of ethanol production from sugarcane. Energ. Policy 36 (6), 2086–2097.

Hettinga, W.G., Junginger, H.M., Dekker, S.C., Hoogwijk, M., McAloon, A.J., Hicks, K.B., 2009. Understanding the reductions in US corn ethanol production costs: an experience curve approach. Energ. Policy 37, 109–203.

Holmes, M.J., Otero, J., Panagiotidis, T., 2013. On the dynamics of gasoline market integration in the United States: evidence from a pair-wise approach. Energ. Econ. 36, 503–510.

Høyer, K.G., 2008. The history of alternative fuels in transportation: the case of electric and hybrid cars. Util. Policy 16 (2), 63–71.

Jacobsson, S., Johnson, A., 2000. The diffusion of renewable energy technology: an analytical framework and key issues for research. Fuel Energy Abst. 28, 624–640.

Junginger, M., Goh, C.S., Faaij, A., 2013. International Bioenergy Trade: History, Status & Outlook on Securing Sustainable Bioenergy Supply, Demand and Markets, vol. 52. Springer Science & Business Media.

Kaup, F., Selbmann, K., 2013. The seesaw of Germany's biofuel policy – tracing the evolvement to its current state. Energ. Policy 62, 513–521.

Kim, J.D., Rahimi, M., 2014. Future energy loads for a large-scale adoption of electric vehicles in the city of Los Angeles: impacts on greenhouse gas (GHG) emissions. Energ. Policy 73 (2014), 620–630.

Lammers, P., Rosillo-Calle, F., Pelkmans, L., Hamelinck, C., 2014. Developments in international liquid biofuel trade International Bioenergy Trade. Springer, Netherlands.17–40.

Lima, M.G.B., Gupta, J., 2013. The policy context of biofuels: a case of non-governance at the global level? Global Environ. Politics 13 (2), 46–64.

Meyer, P.J., 2014. Brazil: political and economic situation and US relations. Current Politics and Economics of South and Central America 7 (1), 1–38.

Mussatto, S.I., Dragone, G., Guimarães, P.M.R., Silva, J.P.A., Carneiro, L.M., Roberto, I.C., et al., 2010. Technological trends, global market, and challenges of bio-ethanol production. Biotechnology Advances 28, 817–830.

Nygren, E., Aleklett, K., Höök, M., 2009. Aviation fuel and future oil production scenarios. Energ. Policy 37 (10), 4003–4010.

OECD/FAO, 2015. OECD-FAO Agricultural Outlook 2015. OECD Publishing, Paris. http://dx.doi. org/10.1787/agr_outlook-2015-en.

Pacini, H., 2015. The Development of Bioethanol Markets under Sustainability Requirements (Doctoral dissertation). KTH Royal Institute of Technology.

Raele, R., Boaventura, J.M.G., Fischmann, A.A., Sarturi, G., 2014. Scenarios for the second generation ethanol in Brazil. Technol. Forecast. Soc. Change 87, 205–223.

RFA (Renewable Fuel Association), Industry Statistics, 2015. Available from: <http://www .ethanolrfa.org/pages/statistics>.

Roehrkasten, S., 2015. Global Governance on Renewable Energy: Contrasting the Ideas of the German and the Brazilian Governments. Springer.

Salles-Filho, S., Bin, A., Drummond, P., Ferro, A.F., Corder, S., Lemos, P., 2015. Innovation in the Brazilian Bioethanol Sector: questioning leadership. In: Globelics International Conference, 13, Cuba.

Searle, S., Sanchez, F.P., Malins, C., German, J., 2014. Technical barriers to the consumption of higher blends of ethanol. ICCT, 1–36.

Sims, R.E.H., Mabee, W., Saddler, J.N., Taylor, M., 2010. An overview of second generation biofuel technologies. Bioresour. Technol. 101 (6), 1570–1580.

Sorda, G., Banse, M., Kemfert, C., 2010. An overview of biofuel policies across the world. Energ. Policy 38 (11), 6977–6988.

Timilsina, G.R., Chisari, O.O., Romero, C.A., 2013. Economy-wide impacts of biofuels in Argentina. Energ. Policy 55, 636–647.

Valencia, M.J., Cardona, C.A., 2014. The Colombian biofuel supply chains: the assessment of current and promising scenarios based on environmental goals. Energ. Policy 67, 232–242.

Van Meijl, H., Smeetsa, E., Zilbermanb, D., 2015. Bioenergy economics and policies. In: Souza, G.M. Victoria, R. Joly, C. Verdade, L. (Eds.), Bioenergy & Sustainability: Bridging the Gaps, vol. 72 SCOPE, Paris, pp. 779. ISBN 978-2-9545557-0-6).

Von Bremen, L., Schmoltzi, M., 1986. Economics and politics of the ethanol market. Trends Biotechnol. 4 (1), 16–23.

Walls, W.D., Rusco, F., Kendix, M., 2011. Biofuels policy and the US market for motor fuels: empirical analysis of ethanol splashing. Energ. Policy 39 (7), 3999–4006.

Conclusions: Futures of Bioethanol—Main Findings and Prospects

S. Salles-Filho

This chapter has the hard task of drawing conclusions about the futures of bioethanol. As the reader most certainly has realized throughout this book, forecasting the futures of bioethanol requires considering many—and sometimes contradictory—criteria. Ethanol, as a fuel, is embedded in an environment full of uncertainties. It is part of the energy strategy of nations and (large) companies. It is, by itself, a matter of discussing the global status quo. When we started writing this book, in the middle of 2015, this was already a tough issue. After the conclusions of the 21st Conference of the Parties (COP21) and its main result, the Paris Agreement, the climate change (and decarbonizing human activities) discussion is now moving to a new and more structured path.

At the very moment this chapter is being written, the Paris Agreement is still a pale—yet important—light, giving navigators the general direction where to point the prow. Nobody really knows what will be the consequences for the energy matrix and even less for biofuels. The commitment of the parties to limit the global warming well below 2°C by 2025 is an important drive, yet vague. Countries are just now starting discussions to convert their Intended Nationally Determined Contribution into real policies and measures.

Actually, the Paris Agreement, despite being a true milestone in the climate change discussion, has been received by numerous experts as a good (and important) "letter of intention" rather than as a statement with targets and deadlines that would drive forecasters to define impacts on the several dimensions of climate change—among which the futures of biofuels are placed.

Regardless of the doubts, the mere existence of the Agreement highlights the importance of low-carbon sources of energy. To which extent this event will attract the attention of policymakers and particularly of companies towards biofuels is something we have to wait to know. Investments in biofuels can either accelerate or keep the pace inside the national energy matrixes and in the international trade scenario.

Just as an exercise of possibilities, let us take into consideration the Intended Nationally Determined Contribution (INDC),[1] of say, four countries/regions whose ethanol policies and strategies were analyzed in this book: the United States, the EU, Brazil and China.

The INDC of the EU and its Member States is "committed to a binding target of at least a 40% domestic reduction in greenhouse gas emissions by 2030 compared to 1990, starting before 2020 (…) The EU and its Member States have already reduced their emissions by around 19% on 1990 levels while GDP has grown by more than 44% over the same period." In spite of having nothing clearly stated in its INDC document, emissions from fuels and transportation are explicitly mentioned as a main objective to be achieved as an overall goal.

As for the United States, its INDC "intends to an economy-wide target of reducing its greenhouse gas emissions by 26–28% below its 2005 level in 2025 and to make its best efforts to reduce its emissions by 28% (…) Achieving the 2025 target will require a further emission reduction of 9–11% beyond 'our' 2020 target compared to the 2005 baseline and a substantial acceleration of the 2005–2020 annual pace of reduction, to 2.3–2.8% per year, or an approximate doubling." There is no explicit mention of biofuels in the document nor any discussion about targets in the transportation sector.

The Brazilian document states: "(…) commits to reducing greenhouse gas emissions by 37% below 2005 levels in 2025. Subsequent indicative contribution: reduce greenhouse gas emissions by 43% below 2005 levels in 2030." According to the document "Brazil's current actions in the global effort against climate change represent one of the largest undertakings by any single country to date, having reduced its emissions by 41% in 2012 in relation to 2005 levels." The document shows the following measures of how they plan to achieve that goal: "i) sustainable use of bioenergy; ii) large-scale measures relating to land use change and forests; tripling, to nearly quadrupling the share of zero- and low-carbon energy supply globally by the year 2050." Regarding biofuels, Brazil is more explicit than other countries by declaring that "it will increase the share of sustainable biofuels in the Brazilian energy mix to approximately 18% by 2030, by expanding biofuel consumption, increasing ethanol supply, including advanced biofuels (second generation), and increasing the share of biodiesel in the diesel mix."

Finally, regarding China, its INDC says that the country "has nationally determined its actions up to 2030 as follows: a) to achieve the peaking of carbon

---

1. The INDC reveals the intention of a nation to reduce its greenhouse gas emissions. Once a country signs the Paris Agreement the INDC will become the NDC, which represents a formal compromise of GHG reduction.

dioxide emissions around 2030 and making best efforts to peak early; b) to lower carbon dioxide emissions per unit of GDP by 60% to 65% from the 2005 level; c) to increase the share of non-fossil fuels in primary energy consumption to around 20%; and d) to increase the forest stock volume by around 4.5 billion cubic trees/meters on the 2005 level." Regarding emissions from transportation, the Chinese INDC states: "to develop a green and low-carbon transportation system, optimizing means of transportation, properly allocating public transport resources in cities, giving priority to the development of public transportation and encouraging the development and use of low-carbon and environment-friendly means of transport, such as new energy vehicles and vessels; to improve the quality of gasoline and to promote new types of alternative fuels; to promote the share of public transport in motorized travel in big-and medium-medium to large sized cities reaching 30% by 2020; to promote the development of dedicated transport system for pedestrians and bicycles in cities and to advocate green travel; to accelerate the development of smart transport and green freight transport."

As can be seen, there is no clear definition about what is going to happen in fuel production and transportation. The three largest economies in the world (considering the EU as a unit) do not say relevant things about how they will accomplish the intended goals of the reduction of greenhouse gas (GHG) emissions. However, biofuels are clearly an option, and probably the easiest to be developed in the short term (by 2025), at least in transportation.

This conclusion reinforces the expectations about how far biofuels can go as global commodities. It also underpins the prospects of new forms of biofuels (processes and products), as well as for the advanced ones, mainly those based on the hydrolysis of cellulosic feedstock. One important observation—which goes in the opposite direction—is the fact that the largest economies do present restrictions in terms of land use change, which is not the case for Brazil, as we have seen in this book.

Whatever the effects of the Paris Agreement on the policies' behavior and markets' reactions are, this is certainly the most important novelty in the bioethanol prospective scenario.

Throughout the chapters of this book the reader is faced with a mix of pros and cons about the prospects of bioethanol. One important remark is that bioethanol is currently an important alternative in two countries, having one of them implicitly or explicitly constantly announcing that the first-generation bioethanol will be replaced by advanced biofuels. The biggest producer of bioethanol in the world (the United States) has made it clear that corn will no longer be the main feedstock for ethanol production. In the Brazilian case its INDC points to an increase in biofuel participation in the national energy matrix.

The expansion of advanced biofuels faces the underdeveloped stage of the technological arsenal to transform different sources of cellulosic feedstock into ethanol—or into another biofuel or even valuable chemical molecules. This is probably one of the main problems to be tackled in the near future. Moreover, it is clearly dependent on the level of investments in R&D and in market development.

The overcoming of technical and economical constraints faced by second-generation bioethanol seems to have reached a turning point: research and industrial activities have achieved a level on which heavy investments are needed to make it economically and technically feasible. This discussion is very well detailed in the Griffin, Saville and MacLean chapters of this book. The success of second-generation ethanol (and other uses of cellulosic feedstock) will strongly depend on further investment in technology developments but also on a combination of rearrangements in the agribusiness sector that are not trivial. These rearrangements demand institutional, cultural and, of course, economic changes and represent a major challenge to make celluloses a prime source of energy and raw material.

As the prices of oil are now favorable to keep consumers burning fossil fuels, the main strength of change will probably come from a stricter/more severe/rigorous regulatory framework. Market forces will not be able to transform the reality of fuel consumption based only upon relative prices. This is surely one of the main constraints for the worldwide expansion of ethanol and other biofuels.

Actually, there is a sort of perverse relation between relative prices of fuels and investment in renewables. The more the relative prices point to favor fossil fuels, the less will be the pressure to increase investments in alternative technologies. On the other hand, the more relative prices favor alternative fuels, the more oil and energy companies will profit from it and the less they will work to find alternatives to fossil fuels. As can be seen in Araújo's chapter, large energy and chemical companies are keen, albeit cautious, to develop new alternatives. However, the pace, path and direction they have undertaken are sometimes ambiguous. Besides, as pointed out by Araújo "at present, there is no durable market signal to make the case for a shift of investment into biofuels by these (large) companies. Their profit margins are simply too big in their core businesses."

In the United States—as pointed out in the chapters by Griffin, Saville and MacLean; Souza et al.; and Araújo—since the regulatory mandates now exceed the production capacity for both corn and cellulosic ethanol, disputes have arisen in the legislative arena trying to reduce or even to eliminate the mandates for ethanol. On the other hand, the recently approved Paris Agreement is working against these movements. Araújo adds an important argument: in the United States, the Federal commitment to subsidize oil and gas is much greater than the support for renewables. Those ups and downs are a good example of the uncertainties surrounding bioethanol.

International trade in bioethanol is considered to evolve at a slow pace. The global forecasts made by OECD and FAO that are mentioned in the Lemos and Mesquita chapter argue that no significant increase of the international flow of ethanol is expected by 2024. Some newcomers are expected to appear on the international trade scene, but all, as minor suppliers (less than 1 billion liters).

On the other hand, the potentiality of international trade can also be quite optimistic. For instance, if we take the projections of using crop wastes (straw and stalks), a study mentioned in Araújo's chapter which shows that by replacing 10% of the gasoline demand with cellulosic ethanol would represent a demand of 115 billion liters/year, while, if the available agricultural residues are converted into fuel that would represent a supply of 351 billion liters/year (1.2–3.5 times the global production of bioethanol to date, respectively).

Of course these figures are based on theoretical possibilities but they help to show how great a potentiality there is that can be explored—and how cautious we have to be, but not be seduced by the numbers nor ignore them.

The wall blend—regulatory limits to blend ethanol to gasoline—is also considered an important constraint to the expansion of the ethanol market. Araújo's chapter and Griffin, Saville and MacLean's chapter on the potential of ethanol in the United States discuss this issue. Despite the worldwide success of blending ethanol to gasoline, today three-fourths of ethanol consumption as a biofuel is limited to 10–20% (E10 or E20). In the United States this restriction is relatively more important since this country is the largest producer in the world and should have a more flexible regulation as to expand the possibilities of blending. Brazil is the country that has experienced the most flexible situation on this issue with flex-fuel engines that accept blends of any proportion.

In the EU, the Renewable Energy Directive (RED) and Fuel Quality Directive (FQD) regulate the production and consumption of biofuels. In this region (different from other developed countries/regions) biofuels have evolved much more based on biodiesel rather than on bioethanol, and advanced biofuels have had little growth in the past decade.

Actually, biofuels in Europe showed a trajectory of fast acceleration at the beginning of 2000 but there was an even faster slowdown after 2008. As is pointed out in Harvey and Bharucha's chapter, changes of regulations and policies experienced by the EU after 2008 determined a fall in biodiesel production by one third in 6 years. Pure biodiesel (B100) has practically disappeared. As argued by the authors above "it signaled the death of the 100% biofuel vehicle in Europe, as a present or future prospect." Recently, when the new 2015 rules concerning GHG were issued and up to now, they have been repositioning biofuels in their EU renewable energy policies.

Land use change, both direct and indirect, is another active source of debate that feeds the stop-and-go policies towards biofuels. If this is not an issue in countries such as Brazil, mentioned before, because of the huge availability of land, it certainly is quite important in Europe and other developing countries, such as India and China. Harvey and Bharucha's chapter stresses the importance of LUC in India, particularly regarding the food and energy security concerns. Climate constraints are clearly ad hoc in this country compared to other emergent economies and the prospects of renewables are very vague, to say the least. With problems of pollution, food and energy security, India's policy towards renewables is perhaps one of the most uncertain prospects among the parties. A quick look at India's INDC reveals a quite firm position towards a reduction of 25% of GHG emissions of 2005 levels.

India's INDC targets, wind and solar energy, are stated as priorities. However, according to this document, energy efficiency is the main goal to be achieved, as coal and burned-biomass are and will remain the main energy sources. As stated in this document, the critical objective is to improve efficiency while reducing emissions proportionally. As for biofuels, the INDC mentions them particularly with the intention of increasing biodiesel production. Ethanol does not seem to be a clear priority to this country. As the chapter of Harvey and Bharucha points out "it is clear that no significant policy action towards biofuels (in India) has been driven by neither climate change nor energy security imperatives."

In the global scenario, it is also imperative to discuss China and its perspectives towards renewables and particularly biofuels. Two chapters of this book discuss China: Harvey and Bharucha and Huaibin Lu. Despite the different perspectives, both chapters agree that China is not heading its energy policy towards biofuels.

Harvey and Bharucha emphasize the trilemma issue in China by stressing the problems related to the per capita availability of agricultural land (it is the lowest when compared to India, the United States and Brazil). Besides this evident and critical problem, the increasing demand for food, energy and water, and the shocking increase in air and water pollution has shaped policies in more careful, and based on necessity, ways. According to these authors, "(…) in marked contrast to the huge progress in other forms of renewable energy, including bioenergy, biofuel consumption in China appears doomed to substantially miss its current targets, largely as a consequence to some systemic constraints to domestic production from China's land and water resources."

In Huaibin Lu's chapter about China, he shows impressive figures that reinforce the worries/concerns pointed out by Mark Harvey and Zareen Bharucha. Just to start, the total number of civilian automobiles in China is now exceeding 150 million and it is expected to double by 2025. In cities like Beijing, 20 out of 100 households own a car. As a consequence, the related demand for gasoline is increasing accordingly. Actually, all energy consumption is increasing. As shown by Lu, recent policies have pointed to the exploration of shale gas and oil, which is a huge advantage as China has the largest reserve in the world.

Despite the growing importance of fossil sources—particularly of shale and oil—China started its bioethanol program at the beginning of the year 2000 and up to now has seven ethanol plants; one fed with celluloses and the others with grains, sorghum and cassava. Altogether, these mills account for around 2.2 billion liters of bioethanol. Despite the increasing volume of production this is still a minor source of fuel in this country. If the internal target of blending 20% of ethanol with gasoline is achieved, it would be necessary to increase 10-fold the present production of bioethanol to around 20 billion liters. Apart from the technical problems pointed out by the author, the main problem seems to be the land use and food production. At the beginning of 2000, internal policies were clearly pro ethanol, indeed, with subsidies and incentives to build plants and to adopt gasoline–ethanol blends. In the past 5 years, China's central government has cut subsidies and other incentives, making producers and consumers move away from ethanol.

At the same time the Chinese government is stimulating other cleaner sources such as gas, biodiesel, methanol fuel and electric cars, all of which are more subsidized than bioethanol. Recent policies have not been directed to ethanol; instead, electric cars and gas-fuel have received much more attention. In Lu's words: "The future fuel ethanol development and utilization are gradually being left for commercial operations and market competition except for the government's controls of grain-based fuel ethanol investment."

He however reminds us that nothing is yet defined and that China counts on huge areas of land that are inappropriate for grain cultivation but able to produce other feedstock to make bioethanol (cassava, sweet potato, sorghum), not to mention the potential of cellulosic ethanol. China has built eight experimental second-generation refineries so far. Last, but not least, Lu calls attention to the potential of China in becoming a major importer of ethanol, which will depend on several circumstances.

Now, turning to Africa, Sergio C. Trindade's chapter shows a quasi-optimistic view of the potential of bioethanol—as to biofuels in general—on that continent. The author starts the chapter stating "the long-term future of biofuels in the world lies in Africa because of its geographical location, resource endowment and increasing energy service needs which are driven by development and population growth. But, in the short term, there has been limited market penetration of biofuels, ethanol and biodiesel in Africa's energy systems."

Although several mandatory or "volunteer mandates" of blending ethanol with gasoline can be found in different African countries, most are not properly working. In Kenya, an early adopter of ethanol during the 1980s, only recently have investments been made in order to explore the "ethanol-as-a-fuel" market. In Bénin the initiatives are even more recent. In this country only minor initiatives have been undertaken. African countries are potential candidates to improve ethanol production, but most of them would have to start from scratch. The initial movement will probably come from policies and regulations, particularly those defining reliable mandates to blending ethanol and gasoline. The trilemma in this continent is not as clear as it is in Europe or in China and India. There is agricultural land, an increasing demand for energy and food, and available technology. However, the political and economic conditions as well as the different development levels blur the scene, making it quite improbable that any of the countries will build a bioethanol market any time soon. Here the potential is still further from the real world than in any of the countries discussed in the other chapters.

The Brazilian case is the most explored in this book. This is the only country that really looks at bioethanol as a complete alternative to fossil fuels, particularly in transportation. It is also the largest producer of ethanol from sugarcane in the world and has a quite important and large industry based on feedstock with hundreds of thousands of employees, huge regions completely planted with sugarcane, and sugar and ethanol refineries and a whole regional economy based on the so-called, sucrose-energy sector. In Brazil and in the United States, the bioethanol industry can be analyzed through the lens of "innovation systems." It is particularly relevant to discuss the futures of bioethanol because this approach comes with technological perspectives of new products and processes and their potential.

As discussed in this book, bioethanol, around the world, has not crossed the thresholds of commoditization although its obvious importance as an alternative to fossil fuels and it has top position in the "league of biofuels" (to use the expression employed by Mark Harvey and Zareen Bharucha in their chapter). Technological innovation is, for certain, among the main variables to be monitored when the issue is forecasting the futures of bioethanol.

The Brazilian and American bioethanol systems of innovation are much more structured than in any other country. Technological developments are in the sights of the main participants of both innovation systems, but with different levels of investments in R&D and in the necessary complementary assets to convert technology in innovation. As the chapter of Salles-Filho et al. about innovation in the ethanol sector in Brazil shows, although relevant investments have been made and important capabilities have been built in that country in the past 20 years, Brazil has not developed proprietary technologies strong enough to set particular technological trajectories.

If we take just first-generation bioethanol, technologies are widely available, whether in Brazil or in the United States. In the Brazilian case, the tradition and ability of producing new sugarcane varieties and in building refineries that are able to produce ethanol and/or sugar—and more recently, to produce electricity from biomass is, of course, an important asset, and the technology is widely diffused. In recent years, the emergence of private companies involved in genetic breeding and in second generation has changed only slightly, but there are no critical technologies that could radically increase the performance of the first generation nor any to promote a technological jump towards the second generation. Despite the tradition and presence of highly qualified research centers and researchers in Brazil, ethanol production has taken an incremental technological path.

The same can be said for the first generation in the United States. This country has made an impressive effort to cope with the mandates of blending ethanol to gasoline and in a few years became the main producer of bioethanol, surpassing Brazil quite easily. From 2006 to 2012 the United States tripled its production of bioethanol, while Brazil remained more or less at the same production level. To give an idea of how impressive this evolution was, Brazil was the leader in ethanol production until 2005; from 2006 on, the United States not only took over the number 1 position, but more than doubled that of Brazil. Technology for the first generation was not an obstacle at all.

Looking at second-generation bioethanol, the situation between these two countries is similar in some aspects and different in others. The similarities are mostly reserved for the declared intention to develop E2G and the existing policies and mechanisms towards cellulosic ethanol. The differences are mostly in the coordination of policies and in the volumes of investments in technology.

In the chapter authored by José Maria F.J. da Silveira and colleagues, there is a clear demonstration of how ahead the United States and China are in terms of technological patents and networks of scientific and technological collaboration. The authors show a scenario of leadership in the science and technology domain of second generation headed by those two countries. Brazil, in spite of its scientific and technological tradition in first generation, is lagging behind in the race for second generation.

While Brazil's policies towards ethanol have been contradictory in the past 5 years, in the United States there is a more coordinated and clear governance. For instance, the expansion of the internal market in Brazil has been precluded because of the fuels price policy imposed by the government which favors gasoline consumption. In a country where flex-fuel engines have grown fast, customers *do* prefer to pay less to fill the tank. Flex-fuel technology makes fuels compete among themselves. There is a choice, and the choice is mainly based on one's pocket.

Comparing the technology systems of innovation in Brazil and the United States, Souza et al., in their chapter, argue that the United States presents higher levels of investment in R&D, in both the first and second generations. Most of the second-generation technologies now being scaled up in Brazil are imported, while in the United States they are local. Another difference relates to the more intense emergence of start-up companies in the United States. Put altogether, the situation points to a scenario in which the United States will probably produce commercial E2G sooner and in a higher volume compared to Brazil. The substitution of the first generation for the second generation seems to be more likely to happen in the United States since Brazil is already behind in the race for E2G, as shown by José Maria F.J. da Silveira and colleagues and other authors in this book.

If technology of cellulosic ethanol succeeds in the US transportation system it will probably be diffused to other countries sooner rather than later. Cellulose is a global feedstock and the more technologies are developed, the more probable they will be adopted in other countries. Agricultural wastes such as straw and stalks are potential sources for producing E2G. There are considerable difficulties in scaling up this kind of feedstock, particularly because cellulosic material does vary in its composition and capacity to feed ethanol refineries. It is more likely—as mentioned in several chapters of this book—that to be successful it is necessary to have regular and standardized cellulosic feedstock.

Luís Augusto Barbosa Cortez in his chapter emphasizes the necessity of Brazil to develop new paths for energy and food production. These new paths should be modulated by sustainability and even the sugarcane production model has to be rethought in order to cope with future requirements of sustainability. Cellulosic-based bioenergy is, for sure, one of the candidates to complement the present sugarcane–sugar–ethanol–electricity model. However, because of that flexible and integrated business model, it will be much more difficult to substitute sugarcane-based ethanol for the cellulosic-based path. This is another reason to believe that in the near future Brazil will keep the present trajectory and remain the main global producer of sugarcane-based bioethanol, even if other technological trajectories appear on the market to compete.

Brazilian success in the sugarcane industry is at the same time, its main strength and its main weakness. The strength can be easily seen throughout history and particularly during the past decades when flex-fuel vehicles came out to dominate the scene of light vehicles; the weakness is due to the lock-in provoked by this trajectory. The advantages of a sugar–ethanol–electricity business model are hard to beat, making the transition to E2G unlikely to happen in the short term. That is why it is better to believe in a complementarity of technological trajectories rather than in substitution.

The main possibility for Brazil to change its plateau of bioethanol production is the expansion of the international market. Other countries are investing in alternatives to fossil fuels (such as biodiesel and electric engines, just to mention the two main direct competitors of ethanol) and also in cellulosic-based bioethanol. The expansion of Brazil's ethanol industry may come about because of one of three circumstances: (1) other countries failing in their attempts to develop better alternatives and increase imports of ethanol; (2) Brazil develops commercial feasibility for E2G; (3) and the last possibility is that Brazil drastically reduces E1G production costs in order to better compete with both fossil and renewable alternatives.

The recent dramatic downward movement of oil prices—from over US$100 to around US$30 (maybe down to 20) in less than 2 years—puts the problem of ethanol costs in a quite exposed situation. First-generation costs, both from corn and sugarcane, are in a threshold of viability when compared to gasoline prices. The decrease in oil prices naturally threatens ethanol's competitiveness. The future of ethanol production now strongly depends on mandates. Its futures—out of the mandates—depend on one of two alternatives: either the first generation proportionally decreases its production costs or the second generation comes with persuasive technical and economic feasibility.

The chapter of Salles-Filho et al. presents concerns about the possibilities of increasing the productivity of sugarcane and ethanol in Brazil, at least, at the pace it would be necessary to face the ups and downs of oil prices and the country's internal policies of price control.

As pointed out by Luís Augusto Barbosa Cortez, in his chapter, a new sustainable agriculture for bioenergy is necessary in Brazil, which has to shift towards "energy cane" and a completely different business model. According to him the present model has already reached its limit.

As argued by Harvey and Bharucha, "(…) in the heyday of biofuel optimism, there was a vision of a new global South-North geopolitical pact for terrestrial transport, with a prospect of '18 Brazils' across the subtropical world providing a substantial energy contribution, both renewable and environmentally beneficial. Less than a decade later, that vision has dimmed, without any significant alternative to the dominance of conventional fuels in transport."

Particularly in Brazil—and also in the United States, although for other reasons—the future of the bioethanol industry is almost totally dependent on the internal market. This is the lifeline for local industry and it seems it will continue being so in the near future. Furthermore, sugar and electricity (as aforementioned) are more than a buffer for the Brazilian sugarcane industry, they are at least as important as ethanol to the industry.

This book has shown the pros and cons of biofuels and particularly of bioethanol. The recent reflux of expectations and beliefs in biofuels do not represent a definitive movement. It is more likely a normal forward–backward thing in the path of energy model transition. The Paris Agreement—and what follows it—will bring about new commitments and measures toward renewables that will probably impact biofuels present and future.

At least four main sources of uncertainties will have to be followed very closely. First, there is a big question mark about how countries will deploy their INDCs into NDCs and which position biofuels and alternative transportation energy sources will take in their internal policies. This is a major uncertainty and can either reverse or reinforce the trends verified in this book (particularly the partial reflux of investments in and the stimulus for renewables in many countries). Secondly, technologies of the second generation are still in a precommercial—not to say development—phase. The feasibility of E2G is the big transformation in ethanol trajectory and can make it a global commodity—depending, of course, on the futures of alternatives and competing sources. Thirdly, fossil fuels are not yet gone and energy efficiency is a main goal among countries. As shown in Huaibin's chapter, in China this seems to be the main path that will be taken before going to renewables. Indeed, "new" sources of fossils, such as shale oil and gas, are changing relative prices and priorities around the world. Besides, the ups and downs of oil prices are still a main reference to public policies and private strategies worldwide. A complementary source of uncertainty is the technological development and the power of changing paradigms. Beyond second-generation technologies no revolutionary technology is on the threshold of commercial feasibility, technological change by its own nature is quite unpredictable and new technological trajectories *do* evolve rapidly depending upon the complexity of capricious events. Electric cars are probably one of the most visible examples: they are on the market, technology is being developed, people are buying them, and relative prices are more or less equilibrated. What is next will be contingent to the same factors the authors have shown in this book.

In general, these uncertainties are actually linked to each other. The movement of one factor changes the movements of the others. This is a characteristic of complex systems where the observed behavior is a matter of evolution through an intended process. It is intentional because the participants make decisions heading toward specific points in the future; it is evolutionary because in a complex system intentions are just what they are: intents.

Index

Printed in the United States
By Bookmasters